Protocols for Studying Pathogenic Fungi

Ritu Pasrija • Deepika Kumari

Protocols for Studying Pathogenic Fungi

Springer

Ritu Pasrija
Department of Biochemistry
Maharshi Dayanand University (MDU)
Rohtak, Haryana, India

Deepika Kumari
Department of Biochemistry
Maharshi Dayanand University (MDU)
Rohtak, Haryana, India

ISBN 978-981-96-7046-8 ISBN 978-981-96-7047-5 (eBook)
https://doi.org/10.1007/978-981-96-7047-5

This Springer imprint is published by the registered company Springer Nature Singapore Pte Ltd.
The registered company address is: 152 Beach Road, #21-01/04 Gateway East, Singapore 189721, Singapore

Dedicated …
To those who dared to dream,
and to those who inspired me to do the same.
For every word, every story,
and every moment of growth—this is for you.

Preface by Ritu Pasrija

It's often said that every story begins with a single spark—a fleeting thought, an unexpected idea, or a quiet moment that demands to be explored. The journey from that initial spark to the final words on a page is never straightforward, and yet it is this very journey that makes writing so profoundly human.

The study of pathogenic fungi is an intricate and dynamic field, encompassing a diverse kingdom that includes both unicellular organisms, like yeasts, and multicellular pathogens, such as molds. These fungi play an outsized role in clinical medicine, agriculture, and environmental science. The complexities of identifying and studying fungal pathogens require not only a deep understanding of their biology but also an adherence to rigorous protocols that ensure accurate results, no matter where in the world those results are obtained.

In this comprehensive volume, *Protocols for Pathogenic Fungi*, the author has made an attempt to compile a collection of protocols that offer precise, reproducible methods for the detection, identification, and study of these diverse and often elusive organisms. The book is a critical resource for clinical microbiologists, researchers, and laboratory technicians, as it provides clear guidelines for handling clinical samples that may harbor yeasts, molds, or dimorphic fungi.

One of the book's standout features is its attention to the variability inherent in working with pathogenic fungi. The fungal kingdom, vast and diverse, presents unique challenges—whether it's the repeatability of results within the same laboratory or across continents. The protocols presented herein are designed with this variability in mind, acknowledging the differences in geography, patient populations, and laboratory settings. The author has gone to great lengths to ensure that these protocols adhere to international standards, including the Clinical and Laboratory Standards Institute (CLSI) guidelines, ensuring both consistency and reliability across the board.

Repeatability is one of the cornerstones of scientific research and clinical diagnostics, and the ability to produce reliable results not only within a single laboratory but across various labs worldwide is a hallmark of good practice. By integrating these standards into the protocols, this book ensures that researchers and clinicians are equipped with the tools necessary to produce consistent and reproducible data, regardless of the setting in which they work.

The idea to compile these protocols was born out of necessity—the need for a standardized approach that takes into account the multiple variables encountered in fungal diagnostics, from sample collection to analysis. The author's thoughtfulness in curating these methods is evident, as they are informed by years of experience and a profound understanding of the challenges and intricacies of fungal diagnostics.

As you delve into the pages that follow, I encourage you to appreciate not only the wealth of practical knowledge contained within this book but also the broader implications it has for global health. In a world where fungal infections are on the rise, this work serves as a vital contribution to improving clinical outcomes, fostering better diagnostic practices, and bridging the gap between laboratories across borders.

The methods outlined in this book will undoubtedly prove indispensable to anyone working in the field of mycology. With each protocol, you will find clear instructions, best practices, and a deeper understanding of how to approach pathogenic fungi in a way that is both scientifically rigorous and clinically relevant.

It is with great pride that I present *Protocols for Pathogenic Fungi*, a work that not only serves the present but will continue to shape the future of fungal diagnostics and research.

Rohtak, Haryana, India Ritu Pasrija

Preface

Studying fungal pathogens, particularly in laboratory settings, demands an intricate balance of technical knowledge, rigorous protocols, and a deep understanding of biosafety. This book is a comprehensive guide to cultivating, handling, and analyzing fungal pathogens in a Biosafety Level 2 (BSL-2) laboratory. The chapters are designed to be a practical reference for researchers, laboratory personnel, and students who wish to deepen their expertise in fungal microbiology, emphasizing safety protocols and laboratory best practices.

Fungal pathogens present significant challenges in terms of treatment and understanding their biological characteristics. From infection control to drug susceptibility testing, understanding the various methods for growing, testing, and analyzing these organisms is essential for improving public health outcomes. The protocols within this book offer detailed insights into working with fungal cultures, assessing their viability, understanding their pathogenic mechanisms, and evaluating potential treatments.

Moreover, this book underscores the importance of adhering to BSL-2 guidelines for safe laboratory practice when handling infectious agents. Each chapter provides detailed methodologies, safety guidelines, troubleshooting tips, and references, ensuring that researchers can execute experiments with confidence, while maintaining safety at every step.

Focusing on practical, hands-on approaches, this work seeks to be an indispensable resource for anyone working with pathogenic fungi in laboratory environments. Whether you're a seasoned researcher or just beginning your journey in microbiology, the information within these pages will help you navigate the complexities of working with fungal pathogens in a BSL-2 setting.

Together, the in vitro and in vivo assays presented in this book offer a complete toolkit for studying fungal pathogens in a laboratory setting. This book is a detailed and systematic guide based on CLSI guidelines for the study and manipulation of fungal pathogens, ensuring safe, effective, and scientifically rigorous approaches for laboratory work involving these microorganisms.

Rohtak, Haryana, India

Ritu Pasrija
Deepika Kumari

Acknowledgments

I begin by expressing my deepest gratitude to the Almighty for His boundless blessings and guidance throughout this journey. His presence has been my constant source of strength, and without His grace, this book would not have come to fruition.

My heartfelt thanks also go to my research scholars, whose continuous support, insightful discussions, and dedication have been integral to completing this work. The numerous discussions, corrections, and constructive feedback have shaped this book, and I truly appreciate their commitment and hard work.

I would like to sincerely thank Sailesh Gochhait for his invaluable moral support during the most challenging moments. His encouragement and belief in me provided the boost I needed to push forward, and I am deeply grateful for his unwavering support.

Last but not least, I dedicate this book to my parents, family, and friends, whose untiring support, love, and sacrifices have been my foundation. They have always stood by me, and their constant encouragement has been a source of inspiration. This achievement is a testament to their unwavering belief in me.

Thank you to all who have been a part of this journey.

Contents

1 Introduction to BSL-2 Facilities 1
1.1 Laboratory Design and Facilities 2
1.1.1 Location and Access 2
1.1.2 Surfaces and Floors 3
1.1.3 Structural Layout 4
1.2 Personal Protective Equipment (PPE) 6
1.2.1 Lab Coats 6
1.2.2 Gloves 8
1.2.3 Eye and Face Protection 10
1.3 Decontamination Procedures 11
1.4 Health and Medical Surveillance 13
1.5 Risk Assessment and Management 14
1.5.1 Steps in Risk Assessment 14
1.5.2 Steps in Risk Management 15
1.6 Documentation and Record Keeping 16
1.7 Training and Education 17
1.7.1 Objectives of Training and Education 17
1.7.2 Key Components of BSL-2 Training Programs 19
1.7.3 Methods of Delivery 20
1.7.4 Evaluation and Certification 22
References 25

2 Handling, Culturing, Preservation and Revival of Fungal Pathogens 27
2.1 Handling Cultures of Fungal Pathogens 28
2.1.1 Biosafety in the Laboratory and Personal Protective Measures 28
2.1.2 Aseptic Techniques 29
2.1.3 Selection and Preparation of Culture Media 29
2.1.4 Culture Maintenance and Subcultures 30
2.1.5 Storage and Transportation of Cultures 31
2.1.6 Disposal of Biosafety Waste 31
2.1.7 Documentation and Records 31
2.1.8 Troubleshooting 31

2.2 Cultivation of Fungal Pathogens 34
2.2.1 Purpose of Cultivation of Fungal Pathogens 35
2.2.2 Isolation of Fungal Pathogens 35
2.2.3 Selection of Growth Media 35
2.2.4 Environmental Conditions for Cultivation. 35
2.2.5 Cultivation Techniques 36
2.2.6 Special Cultivation Requirements 37
2.2.7 Dealing with Contamination 37
2.2.8 Quality Control and Monitoring 37
2.2.9 Challenges and Solutions. 37
2.2.10 Applications of Fungal Cultures 37
2.2.11 Troubleshooting. 38
2.3 Preservation of Fungal Cells 40
2.3.1 Importance of Fungal Cell Preservation 40
2.3.2 Methods for the Preservation of Fungal Cells 41
2.3.3 Challenges in the Preservation of Fungal Cells. 43
2.3.4 Factors Influencing the Success of Preservation 43
2.3.5 Applications of Preserved Fungal Cells. 43
2.3.6 Troubleshooting. 44
2.4 Resuscitation of Fungal Cells 47
2.4.1 Importance of Fungal Cultures Revival. 47
2.4.2 General Steps in the Revival Process. 47
2.4.3 Quality Control 48
2.4.4 Challenges in the Revival of the Fungus 48
2.4.5 Specialized Revival Techniques. 49
2.4.6 Applications of Revived Fungal Cultures 49
2.4.7 Troubleshooting. 49
References. 52

3 Counting Colony Forming Units (CFU) to Estimate Viable Fungal Cells in a Sample 53
3.1 Materials Required. 54
3.1.1 Fungal Sample and Medium 54
3.1.2 Materials for Dilution 54
3.1.3 Plating Materials 54
3.1.4 Sterilization and Safety Accessories 55
3.1.5 Incubation Equipment 55
3.1.6 Counting Equipment 55
3.1.7 Data Recording Tools 55
3.2 Preparations of Media and Buffers 55
3.2.1 Preparation of YEPD Media and Agar (See Chap. 2) . . . 55
3.2.2 Preparation of Phosphate-Buffered Saline (PBS) 55
3.2.3 Preparation of the Saline Solution (0.85%). 56
3.3 Procedure 56
3.3.1 Preparation of the Fungal Culture 56

3.3.2 Serial Dilution 57
3.3.3 Plating Out the Dilutions 57
3.3.4 Incubation 58
3.3.5 Count the Colonies 58
3.3.6 Calculation of CFU/mL 58
3.4 Tips and Troubleshooting 58
3.4.1 Tips 58
3.4.2 Troubleshooting 59

4 Growth Curve Preparation for the Calculation of Fungal Cell Count Over Time 63
4.1 Materials Required 64
4.1.1 Fungal Culture and Growth Media 64
4.1.2 Culture Containers 64
4.1.3 Sampling Tools 64
4.1.4 Measurement Devices 64
4.1.5 Sterile Filtration and Dilution Materials 65
4.1.6 Time and Temperature Control 65
4.1.7 Sterilization and Safety Materials 65
4.1.8 Data Recording and Analysis 65
4.2 Preparation of the Media and Buffers 65
4.2.1 Preparation of YEPD Media (See Chap. 2) 65
4.2.2 Preparation of RPMI-1640 Media (See Chap. 2) 65
4.2.3 Preparation of PBS (See Chap. 3) 65
4.2.4 Preparation of YNB Medium and Agar 65
4.3 Protocol 66
4.3.1 Preparation of the Primary Culture and Application of the (Spotting) Cells to the Plates for Counting the CFU 67
4.3.2 Serial Dilution and Plating for CFU Counting 67
4.3.3 Calculation of CFU/mL 67
4.3.4 Plotting of the Growth Curve 68
4.3.5 Statistical Evaluation 68
4.4 Tips and Troubleshooting 69
4.4.1 Tips 69
4.4.2 Troubleshooting 70

5 Microtiter Assay for Drug Susceptibility Testing of Pathogenic Fungi in Broth 73
5.1 Materials Required 74
5.2 Preparations 74
5.2.1 Preparation of Media 74
5.2.2 Preparation of Drug Stocks 74
5.2.3 Preparation of Fungal Culture 75
5.3 Procedure 76

5.4 Tips and Troubleshooting . . . 77
5.4.1 Tips . . . 77
5.4.2 Troubleshooting . . . 78
References . . . 80

6 Disk Diffusion Assay for Testing the Drug Sensitivity of Pathogenic Fungi in Agar . . . 81
6.1 Materials Required . . . 82
6.1.1 Agar Plates . . . 82
6.1.2 Fungal Culture . . . 82
6.1.3 Antifungal Disks . . . 82
6.1.4 Aids for the Inoculum Preparation . . . 83
6.1.5 Incubation . . . 83
6.1.6 Sterile Forceps or Tweezers . . . 83
6.1.7 Caliper or Ruler . . . 83
6.1.8 Biosafety and Sterilization Equipment . . . 83
6.1.9 Cleaning Agents . . . 83
6.2 Preparations . . . 84
6.2.1 Preparation of Culture Media and Agar (See Chap. 2) . . . 84
6.2.2 Preparation of the Fungal Culture (See Chap. 3) . . . 84
6.2.3 Preparation of the Paper Disks . . . 84
6.2.4 Preparation of Drug Stocks (See Chap. 5) . . . 85
6.3 Procedure . . . 85
6.4 Tips and Troubleshooting . . . 86
6.4.1 Tips . . . 86
6.4.2 Troubleshooting . . . 87
References . . . 90

7 Spot Assay for Comparing Drug Susceptibility of Pathogenic Fungi in Agar . . . 91
7.1 Materials Required . . . 92
7.1.1 Fungal Strains and Growth Media . . . 92
7.1.2 Antifungal Drugs . . . 92
7.1.3 Preparation of the Fungal Suspension . . . 92
7.1.4 Agar Plates . . . 93
7.1.5 Spotting Tools . . . 93
7.1.6 Sterilization and Safety Materials . . . 93
7.1.7 Incubation . . . 93
7.1.8 Tools for Data Recording . . . 93
7.2 Preparation of Media and Buffers . . . 94
7.2.1 Preparation of YEPD Media and Agar (See Chap. 2) . . . 94
7.2.2 Preparation of PBS (See Chap. 3) . . . 94

7.2.3 Preparation of Saline Solution (0.85%) (See Chap. 3) 94
7.2.4 Preparation of Drug Stocks (See Chap. 5). 94
7.3 Protocol 94
7.3.1 Inoculation of the Strains. 94
7.3.2 Serial Dilution 94
7.3.3 Spot the Dilutions onto Agar Plates. 94
7.3.4 Analysis 95
7.4 Tips and Troubleshooting. 96
7.4.1 Tips 96
7.4.2 Troubleshooting. 96

8 Time-Kill Kinetics to Determine the Number of Living Cells Over a Period of Time. 99
8.1 Materials Required. 100
8.1.1 Fungal Culture and Medium 100
8.1.2 Antifungal Agents 100
8.1.3 Materials for Sampling and Dilution 100
8.1.4 Incubation and Growth Monitoring. 101
8.1.5 Sterilization and Safety Materials 101
8.1.6 Counting Tools/Devices. 101
8.1.7 Tools for Data Recording. 101
8.2 Preparation of Media and Buffers 101
8.2.1 Preparation of YEPD Media (See Chap. 2) 101
8.2.2 Preparation of PBS and 0.85% Saline Solution (See Chap. 3) 101
8.2.3 Preparation of Yeast Nitrogen Base Medium (YNB) and Agar (See Chap. 4) 101
8.3 Protocol 102
8.3.1 Preparation of the Primary Culture 102
8.3.2 Setting up the Plate 102
8.3.3 Incubation and Time Points. 102
8.3.4 Dilution Series for Live Cell Counting 102
8.3.5 Counting the Colonies 103
8.3.6 Statistical Evaluation 103
8.4 Tips and Troubleshooting. 104
8.4.1 Tips 104
8.4.2 Troubleshooting. 105

9 Checkerboard Analysis to Calculate the Fractional Inhibitory Concentration (FIC) Index of Pathogenic Fungi for Drug Interaction 109
9.1 Materials Required. 110
9.2 Preparations 110
9.2.1 Preparation of the Culture Media (See Chap. 2) 110
9.2.2 Preparation of the Fungal Culture (See Chap. 3). 110

9.2.3 Preparation of the Drug Stocks (See Chap. 5) 110
9.3 Procedure 110
9.4 Tips and Troubleshooting 113
9.4.1 Tips 113
9.4.2 Troubleshooting 114
References 116

10 Hyphal Induction in Yeasts and Molds 117
10.1 Materials Required 118
10.1.1 Fungal Strains 118
10.1.2 Growth and Induction Media 118
10.1.3 Materials for Environmental Control 119
10.1.4 Sterile Materials for Sample Handling 119
10.1.5 Microscopy Accessories 119
10.1.6 Safety and Aseptic Materials 119
10.1.7 Tools for Time Measurement and Observation 119
10.2 Preparation of Media and Buffers 120
10.2.1 Preparation of YEPD +10% Serum 120
10.2.2 Preparation of RPMI-1640 + 10% Serum 120
10.2.3 Preparation of PBS (See Chap. 3) 121
10.2.4 Preparation of Spider Medium 121
10.3 Protocol 121
10.3.1 Prepare the Yeast Culture 121
10.3.2 Inoculation in Hyphae-Inducing Media 122
10.3.3 Incubation 122
10.3.4 Microscopic Observation 122
10.4 Tips and Troubleshooting 123
10.4.1 Tips 123
10.4.2 Troubleshooting 124
References 126

11 Isolation of Genomic DNA from Pathogenic Fungi 127
11.1 Materials Required 128
11.1.1 Fungal Strains and Culture Media 128
11.1.2 Materials for the Collection of Fungal Cells 128
11.1.3 Reagents for Disrupting the Cell Wall 128
11.1.4 Reagents for DNA Extraction 129
11.1.5 DNA Precipitation and Purification 129
11.1.6 Materials for Quality Control 129
11.1.7 Safety and Aseptic Materials 130
11.2 Preparations 130
11.2.1 Preparation of YEPD Media (See Chap. 2) 130
11.2.2 Preparation of PDB (See Chap. 2) 130
11.2.3 Preparation of the Phenol-Chloroform-Isoamyl Alcohol Solution (PCI) 130
11.2.4 Preparation of the TAE Buffer 131

11.2.5 Preparation of Breaking Buffer 131
11.2.6 Preparation of Urea Buffer 132
11.2.7 Preparation of Lysis Buffer 133
11.2.8 Preparation of CTAB Extraction Buffer (2%) 133
11.2.9 Preparation of Phenol-Chloroform (1:1) 134
11.2.10 Preparation of RNase A Stock Solution 134
11.2.11 Preparation of 0.8% Agarose Gel 135
11.3 DNA Extraction Procedure for *Candida*/*Saccharomyces* Sp. 135
11.4 Procedure for DNA Extraction from *Cryptococcus* Sp. (Fig. 11.2) 136
11.5 Procedure for DNA Extraction from *Fusarium*/*Aspergillus* Sp. 138
11.6 Gel Electrophoresis 138
11.7 Tips and Troubleshooting 140
11.7.1 Tips 140
11.7.2 Troubleshooting 141
References 143

12 Exopolysaccharide Secretion by Pathogenic Fungi 145
12.1 Materials Required 146
12.1.1 Fungal Strains and Growth Media 146
12.1.2 Materials to Stimulate EPS Secretion 146
12.1.3 Materials for Isolation of EPS 147
12.1.4 Materials for the Characterization of EPS 147
12.1.5 Safety and Aseptic Materials 147
12.2 Preparation of Media and Buffers 148
12.2.1 Preparation of YEPD Media (See Chap. 2) 148
12.2.2 Preparation of PBS (See Chap. 3) 148
12.2.3 Preparation of Isotonic Buffer 148
12.3 Protocol for EPS Extraction and Storage 148
12.3.1 Materials Required 148
12.3.2 Procedure 149
12.4 Quantification of EPS After Storage 149
12.4.1 Thawing and Preparations 149
12.4.2 Phenol-Sulfuric Acid Assay 150
12.5 Tips and Troubleshooting 151
12.5.1 Tips 151
12.5.2 Troubleshooting 152

13 Biofilm Formation Test and Determination of the Effect of Drugs on Their Formation 155
13.1 Materials Required 156
13.1.1 Fungal Strains and Growth Media 156
13.1.2 Materials for Cultivation and Assay Setup 157
13.1.3 Materials for Drug Testing 157
13.1.4 Biofilm Staining and Quantification 157
13.1.5 Equipment for Detection 157

13.1.6 Controls (Optional) 158
13.2 Preparations 158
13.2.1 Preparation of YEPD Media (See Chap. 2) 158
13.2.2 Preparation of RPMI-1640 (See Chap. 2) 158
13.2.3 Preparation of PBS (See Chap. 3) 158
13.2.4 Preparation of Saline Solution (0.85%) (See Chap. 3) 158
13.3 Formation of Biofilms and Disruption 158
13.3.1 Ingredients 158
13.3.2 Procedure 158
13.4 Quantification of Biofilms after Crystal Violet Staining and Microscopy 159
13.4.1 Materials Required 159
13.4.2 Procedure 160
13.5 Tips and Troubleshooting 160
13.5.1 Tips 160
13.5.2 Troubleshooting 162
References 164

14 Induction of Capsule Formation in *Cryptococcus* sp. 165
14.1 Materials Required 166
14.1.1 Fungal Strains and Growth Media 166
14.1.2 Environmental Conditions 166
14.1.3 Reagents for Visualization of the Capsules 166
14.1.4 Microscopy and Imaging Techniques 167
14.1.5 Materials for Handling the Cultures 167
14.1.6 Materials for the Quantification of Capsule Formation . . 167
14.1.7 Controls (Optional) 167
14.1.8 Safety and Aseptic Materials 167
14.2 Preparations of Media and Buffers 168
14.2.1 Preparation of YEPD Media (See Chap. 2) 168
14.2.2 Preparation of SDB Media 168
14.2.3 Preparation of PBS (See Chap. 3) 168
14.3 Protocol for Capsule Formation and Analysis [1–3] 168
14.3.1 Ingredients 168
14.3.2 Procedure 169
14.3.3 Quantification of Capsule Production 169
14.4 Tips and Troubleshooting 172
14.4.1 Tips 172
14.4.2 Troubleshooting 172
References 176

15 Melanin Formation in *Cryptococcus* sp. 177
15.1 Materials Required 178
15.1.1 Fungal Strains and Culture Media 178
15.1.2 Laboratory Consumables 178

15.1.3 Microscopy and Imaging Techniques 179
15.1.4 Environmental Conditions 179
15.1.5 Controls 179
15.1.6 Safety and Aseptic Materials 179
15.2 Preparations of Media and Buffers 179
15.2.1 Preparation of YEPD Media (See Chap. 2) 179
15.2.2 Preparation of L-DOPA Stock Solution. 179
15.2.3 Preparation of Minimal Media (MM) [1] 180
15.2.4 Preparation of PBS (See Chap. 3) 181
15.3 Protocol for Melanin Formation 181
15.3.1 Materials Required. 181
15.3.2 Procedure 181
15.4 Protocol for Melanin Quantification 182
15.4.1 Quantification of Melanin with Adobe Photoshop CS3 and Microscopy 182
15.4.2 Microscopic Observation for Melanin Localization 183
15.5 Tips and Troubleshooting. 184
15.5.1 Tips 184
15.5.2 Troubleshooting 185
References. 188

16 Urease Assay for *Cryptococcus* sp. 189
16.1 Materials Required. 190
16.1.1 Fungal Strains and Growth Media 190
16.1.2 Reagents for Detection of Urease Activity 190
16.1.3 Laboratory Consumables 190
16.1.4 Instruments 191
16.1.5 Controls 191
16.1.6 Environmental and Safety Materials 191
16.2 Preparations of Media and Buffers 191
16.2.1 Preparation of the YEPD Media (See Chap. 2) 191
16.2.2 Preparation of 0.85% Saline Solution (See Chap. 3). . . . 191
16.2.3 Preparation of PBS (See Chap. 3) 191
16.2.4 Preparation of Christensen's Urea Agar Plates 191
16.3 Protocol 192
16.3.1 Materials Required. 192
16.3.2 Procedure 192
16.4 Tips and Troubleshooting. 193
16.4.1 Tips 193
16.4.2 Troubleshooting 195
References. 197

17 Phospholipase Assay for the Detection of Virulence of Pathogenic Fungi 199
17.1 Materials Required 200
17.1.1 Fungal Strains and Growth Media 200
17.1.2 Specialized Phospholipase-Inducing Media 200
17.1.3 Reagents and Chemicals 200
17.1.4 Laboratory Consumables 201
17.1.5 Instruments 201
17.1.6 Controls 201
17.1.7 Safety and Aseptic Materials 201
17.2 Preparations 201
17.2.1 Preparation of YEPD Media (See Chap. 2) 201
17.2.2 Preparation of PBS (See Chap. 3) 201
17.2.3 Preparation of Egg Yolk Agar Plates 201
17.3 Protocol 202
17.4 Tips and Troubleshooting 204
17.4.1 Tips 204
17.4.2 Troubleshooting 205

18 Hydrophobicity Test for Virulence of Pathogenic Fungi 209
18.1 Materials Required 210
18.1.1 Fungal Strains and Growth Media 210
18.1.2 Reagents for the Test 210
18.1.3 Laboratory Consumables 210
18.1.4 Equipment 211
18.1.5 Controls 211
18.1.6 Environmental and Safety Materials 211
18.2 Preparations 211
18.2.1 Preparation of YEPD Media (See Chap. 2) 211
18.2.2 Preparation of PBS (See Chap. 3) 211
18.3 Protocol [1] 211
18.3.1 Preparation of Yeast Cells 211
18.3.2 MATH Assay Procedure 212
18.3.3 Measuring the Optical Density 212
18.4 Tips and Troubleshooting 213
18.4.1 Tips 213
18.4.2 Troubleshooting 214
References 216

19 Adhesion of Fungal Cells to Buccal Epithelial Cells 217
19.1 Materials Required 218
19.1.1 Biological Materials 218
19.1.2 Staining Agents and Reagents 218
19.1.3 Laboratory Consumables 219
19.1.4 Instruments 219

19.2 Preparation of Media and Buffers 219
19.2.1 Preparation of YEPD Media (See Chap. 2) 219
19.2.2 Preparation of RPMI-1640 (See Chap. 2) 219
19.2.3 Preparation of PBS (See Chap. 3) 219
19.2.4 Preparation of Spider Media (See Chap. 8) 219
19.3 Protocol [1] 219
19.4 Tips and Troubleshooting 221
19.4.1 Tips 221
19.4.2 Troubleshooting 222
References 224

20 Sterols Extraction, Analysis and Quantification in Pathogenic Fungi 225
20.1 Material Required 226
20.1.1 Biological Materials 226
20.1.2 Chemicals and Reagents 226
20.1.3 Equipment 227
20.1.4 Safety Materials 227
20.2 Preparation of Media and Buffers 227
20.2.1 Preparation of YEPD Media (See Chap. 2) 227
20.2.2 Preparation of 25% Alcoholic KOH Solution 227
20.3 Protocol 228
20.4 Tips and Troubleshooting 229
20.4.1 Tips 229
20.4.2 Troubleshooting 230
References 233

21 Assessing Mitochondrial Fitness by Assessing Microbial Growth on Non-fermentable Carbon Sources 235
21.1 Materials Required 236
21.1.1 Biological Materials 236
21.1.2 Chemicals and Reagents 236
21.1.3 Laboratory Consumables 237
21.1.4 Equipment 237
21.2 Preparation of Media and Buffers 237
21.2.1 Preparation of Yeast Extract Peptone Dextrose (YEPD) Media (See Chap. 2) 237
21.2.2 Preparation of PBS (See Chap. 3) 237
21.2.3 Preparation of 0.85% Saline (See Chap. 3) 237
21.2.4 Preparation of YNB with NFCs 237
21.3 Protocol 238
21.4 Tips and Troubleshooting 239
21.4.1 Tips 239
21.4.2 Troubleshooting 241
References 243

22 In Vivo Assay Using *Caenorhabditis elegans* 245
22.1 Material Required 246
22.1.1 Biological Materials 246
22.1.2 Chemicals and Reagents 247
22.1.3 Laboratory Consumables 247
22.1.4 Equipment 247
22.2 Preparation of Media and Buffers for the Maintenance of *C. elegans* 247
22.2.1 Preparation of NGM Plates 247
22.2.2 Preparation of Bacterial Food Source 249
22.2.3 Preparation of M9 Buffer for *C. elegans* 249
22.2.4 Seeding NGM Plates with *E. coli* OP50 250
22.2.5 Preparation of Luria-Bertani (LB) Broth 251
22.2.6 Preparation of BHI Agar Plates 251
22.2.7 Preparation of YEPD Media (See Chap. 2) 252
22.3 Synchronization and Staging of the Cultures 252
22.3.1 Transferring Worms Grown on NGM Plates 252
22.3.2 Embryo Isolation 253
22.3.3 Life Cycle from Egg to Egg-Bearing Adult 255
22.4 Picking out *C. elegans* 255
22.5 Infection Model [2] 256
22.5.1 Preparation of Fungal Culture 257
22.5.2 Infection Setup 258
22.5.3 Monitoring the Infection 258
22.5.4 Imaging the Worms 258
22.5.5 To Identify Gastrointestinal Abnormalities 258
22.5.6 Data Analysis 259
22.6 Guidelines for Freezing Worms for Prolonged Storage 260
22.7 Tips and Troubleshooting 260
References 262

About the Authors

Ritu Pasrija is an esteemed Associate Professor in the Department of Biochemistry at Maharshi Dayanand University (MDU), where she has made significant contributions to the field of biochemistry and microbiology. With over 20 years of experience in the study of drug resistance, her research primarily focuses on pathogenic fungi, including *Candida* sp., *Cryptococcus*, and *Fusarium*.

Dr. Pasrija completed her PhD at Jawaharlal Nehru University, one of India's leading institutions, where she deepened her understanding of molecular biology and microbial pathogenesis. Before her academic career, she worked at Ranbaxy Pharmaceuticals, where she gained valuable industry experience, particularly in pharmaceutical research and development. Her research expertise lies in the mechanisms of drug resistance, with a particular interest in antifungal resistance. Over the years, she has undertaken several research projects funded by prestigious agencies such as the University Grants Commission (UGC), the Department of Science and Technology (DST), and the Department of Biotechnology (DBT). These projects have furthered our understanding of the pathogenicity and resistance mechanisms of fungi that are responsible for various infections, particularly in immunocompromised individuals. In addition to her research work, she has published numerous research papers, review articles, and book chapters in reputable journals, contributing extensively to the scientific community. Her dedication to scientific excellence has been recognized by her peers, and she was honored with the Research Excellence Award at Maharshi Dayanand University. Her continued work in the field of drug resistance, particularly in the context of fungal pathogens, remains at the forefront of efforts to develop more effective treatments for infections caused by resistant organisms.

Deepika Kumari is a dedicated PhD student in the Department of Biochemistry at Maharshi Dayanand University, Rohtak, India. In recognition of her exceptional academic potential and in support of her doctoral research, she has been awarded the prestigious Junior Research Fellowship (JRF) and Senior Research Fellowship (SRF) by the Council of Scientific & Industrial Research (CSIR), India. She has made significant contributions to the scientific community and has published five research papers in high-impact, peer-reviewed journals. She has also authored six book chapters and a comprehensive review article, demonstrating her expertise and tireless commitment to advancing the field of mycology. Her research focuses on the biology of fungi and their wider biochemical implications, demonstrating her deep passion for the subject and her potential to significantly contribute to the field. Her work is a testament to her dedication and growing stature as a scientist in biochemistry and mycology.

Abbreviations

°C	Celsius
µg	Microgram
µL	Microliter
µm	Micrometer
A_0	Initial Optical Density
A_1	Optical Density of the aqueous phase after phase separation
A260/A280	Absorbance ratio used to assess DNA purity (measured at 260 nm and 280 nm)
AIDS	Acquired Immunodeficiency Syndrome
ATP	Adenosine Triphosphate
BCA	Biosafety Cabinet Area
BECs	Buccal Epithelial Cells
BHA	Brain Heart Infusion Agar
BLS	Basic Life Support
BMT	Biological Material Transportation
BSA	Bovine Serum Albumin
BSC	Biological Safety Cabinet
BSL-2	Biosafety Level 2
C. albicans	*Candida albicans*
C. elegans	*Caenorhabditis elegans*
C. gattii	*Cryptococcus gattii*
C. neoformans	*Cryptococcus neoformans*
$CaCl_2$	Calcium Chloride
CDC	Centers for Disease Control and Prevention
CFCS	Cell-Free Culture Supernatant
CFU	Colony Forming Unit
CFW	Calcofluor White
CLIA	Clinical Laboratory Improvement Amendments
CLSI	Clinical and Laboratory Standards Institute
CLSM	Confocal Laser Scanning Microscope
cm	Centimeter
CO_2	Carbon Dioxide
CS3	Adobe Photoshop CS3
CSH	Cell Surface Hydrophobicity

CTAB	Cetyltrimethylammonium Bromide
DAPI	4′,6-Diamidino-2-Phenylindole
DHE	Dehydroergosterol
DMEM	Dulbecco's Modified Eagle Medium
DMSO	Dimethyl Sulfoxide
DNA	Deoxyribonucleic Acid
DTT	Dithiothreitol
E. coli	*Escherichia coli*
EDTA	Ethylenediaminetetraacetic Acid
ELISA	Enzyme-Linked Immunosorbent Assay
EPA	Environmental Protection Agency
EPS	Exopolysaccharide
EtBr	Ethidium Bromide
EUCAST	European Committee on Antimicrobial Susceptibility Testing
FBS	Fetal Bovine Serum
FDA	Food and Drug Administration
FIC	Fractional Inhibitory Concentration
FIC Index	Fractional Inhibitory Concentration Index
FITC	Fluorescein Isothiocyanate
FTIR	Fourier Transform Infrared Spectroscopy
FUdR	5-Fluoro-2′-deoxyuridine
g	Gram
g	Gravitational force, used in centrifugation
GalXM	Galactoxylomannan
GC	Gas Chromatography
GFP	Green Fluorescent Protein
GXM	Glucuronoxylomannan
h	Hour
H_2O_2	Hydrogen Peroxide
H_2SO_4	Sulfuric Acid
HBV	Hepatitis B Virus
HCl	Hydrochloric Acid
HEPA	High-Efficiency Particulate Air
HIV	Human Immunodeficiency Virus
HPLC	High-Performance Liquid Chromatography
HRSA	Health Resources and Services Administration
HVAC	Heating, Ventilation, and Air Conditioning
HWP1	Hyphal Wall Protein 1
KCl	Potassium chloride
KH_2PO_4	Monopotassium Phosphate
KOH	Potassium Hydroxide
L	Liter
L-DOPA	L-3,4-Dihydroxyphenylalanine
LB	Luria-Bertani (medium)

LC-MS/MS	Liquid Chromatography-Mass Spectrometry/Mass Spectrometry
LDH	Lactate Dehydrogenase
LIMS	Laboratory Information Management System
LM	Light Microscope
LT50	Lethal Time for 50% of worms to die
M	Molar (concentration)
MATH	Microbial Adhesion to Hydrocarbons
MCF	Mycelial Colony Formation
McFarland	McFarland Standard
MEA	Malt Extract Agar
MFC	Minimum Fungicidal Concentration
MFI	Mean Fluorescence Intensity
mg	Milligram
$MgSO_4$	Magnesium Sulfate
MIC	Minimal Inhibitory Concentration
mL	Milliliter
mM	Millimolar
MM	Minimal Medium
mmHg	Millimeters of Mercury
MOPS	Morpholinepropanesulfonic Acid
MQ	Milli-Q
MSDS	Material Safety Data Sheets
mV	Millivolt
N95	Respirator Mask (N95 standard)
Na_2HPO_4	Disodium Hydrogen Phosphate
$Na_2HPO_4{\cdot}2H_2O$	Disodium Hydrogen Phosphate Dihydrate
NaCl	Sodium chloride
NaN_3	Sodium Azide
NaOH	Sodium Hydroxide
NCCLS	National Committee for Clinical Laboratory Standards
NFC	Non-Fermentable Carbon Source
NGM	Nematode Growth Medium
NSA	Niger Seed Agar
OD_{600}	Optical density at 600 nm
OP50	*Escherichia coli* OP50 strain
OSHA	Occupational Safety and Health Administration
PBS	Phosphate-buffered saline
PCI	Phenol-Chloroform-Isoamyl Alcohol
PCR	Polymerase Chain Reaction
PDA	Potato Dextrose Agar
PEP	Post-Exposure Prophylaxis
PFA	Personal Protection and First Aid
pH	Power of Hydrogen
PPE	Personal Protective Equipment

Ppm	Parts Per Million
PVP	Polyvinylpyrrolidone
Pz	Phospholipase Production Zone
RG	Risk Group
RMP	Risk Management Plan
RNA	Ribonucleic Acid
RNase A	Ribonuclease A
ROS	Reactive Oxygen Species
RPE	Respiratory Protective Equipment
rpm	Revolutions Per Minute (used for centrifuge speed)
RPMI	Roswell Park Memorial Institute
RT	Room Temperature
RT-PCR	Reverse Transcription Polymerase Chain Reaction
SAA	Segregated Area for Autoclaving
SARS-CoV-2	Severe Acute Respiratory Syndrome Coronavirus 2
SC	Synthetic Complete Medium
SC5314	A specific strain of *Candida albicans*
SD	Standard Deviation
SDA	Sabouraud Dextrose Agar
SDB	Sabouraud Dextrose Broth
SDS	Sodium Dodecyl Sulfate
SEM	Scanning Electron Microscopy
SOPs	Standard Operating Procedures
sp	Species
SPSS	Statistical Package for the Social Sciences
SWI	Sharps Waste Incineration
TAE	Tris-Acetate-EDTA
TBE	Tris-Borate-EDTA
TE	Tris-EDTA
TSA	Tryptic Soy Agar
TWA	Time-Weighted Average
U/mL	Units per Milliliter
UV	Ultraviolet
UV-VIS	Ultraviolet-Visible (Spectrophotometry)
V/V	Volume/Volume
WHO	World Health Organization
WT	Wild-Type
YEPD	Yeast Extract Peptone Dextrose
YNB	Yeast Nitrogen Base

Introduction to BSL-2 Facilities

1

Abstract

Biosafety Level 2 (BSL-2) facilities are specialized laboratories that ensure the safe handling of medium-risk biological agents, mainly risk group (RG) 2 organisms that pose a risk to personnel and the environment. These laboratories are essential to advance research, diagnostics and biotechnological applications involving pathogens associated with human disease. Key features of BSL-2 facilities include restricted access for authorized personnel, mandatory use of personal protective equipment (PPE) such as lab coats, gloves and face shields, and containment measures such as biosafety cabinets (Class II) to minimize exposure to infectious agents. Laboratory ventilation systems are designed to direct airflow from clean to contaminated areas to prevent the release of pathogens to the outside environment. Safety equipment, including autoclaves, eyewash stations and safety showers, support rapid response to potential incidents (Biosafety Manuals, ICMR Publications, 2024). These laboratories play a central role in microbiology, virology, diagnostic testing, teaching and provide essential infrastructure for advancing scientific and public health goals. Compared to other biosafety levels, BSL-2 laboratories strike a balance between operational efficiency and rigorous biosafety protocols. The robust security practices combined with a controlled environment make BSL-2 facilities an indispensable part of modern biological research. They protect laboratory personnel, the environment and public health while enabling breakthrough discoveries in the life sciences. The Department of Biotechnology, India has set specific guidelines for BSL-2 laboratories, which include design specifications, required safety equipment, operational practices and safety protocols (Guidelines, DBT BSL2 and BSL3 and Certification of BSL3 facility 2020). In addition, these laboratories must adhere to international safety standards set by organizations such as the World Health Organization (WHO) and the Centers for Disease Control and Prevention (CDC). The guidelines ensure that facilities can safely work with genetically

R. Pasrija, D. Kumari, *Protocols for Studying Pathogenic Fungi*,
https://doi.org/10.1007/978-981-96-7047-5_1

modified organisms, handle biological material and conduct research that could use RG 2 pathogens. The most important points of the BSL2 guidelines are summarized in this chapter.

Keywords
BSL2 · Risk group 2 · Laboratory · Biosafety protocols · PPE

BSL-2 Guidelines BSL-2 is the standard level for laboratories handling medium-risk pathogens, including many fungal pathogens (*Candida* sp., *Aspergillus* sp., etc.). These organisms pose a risk by inhalation, ingestion or contact with mucous membranes or broken skin, but are not normally transmitted by incidental casual contact. BSL-2 facilities ensure the safety of workers and the protection of the environment. The Department of Biotechnology, under the Ministry of Science & Technology, Government of India has formulated the recommendations and guidelines for biosafety. Among these different sets of documents, the "Guidelines and Handbook for Institutional Biosafety Committees" (IBSC) revised edition in May 2011 is a manual for setting up BSL-2 laboratories, setting up IBSC at the institutional level, its registration, review meetings, and approval. Some key recommendations related to setting up BSL-2 facilities are discussed in the following sections-.

1.1 Laboratory Design and Facilities

The architectural and functional design of these facilities focuses on minimizing risks to personnel, the environment, and the public while maintaining operational flexibility for scientific activities. The following are some key considerations for laboratory and facilities design.

1.1.1 Location and Access

Appropriate location and controlled access are critical to safety and compliance. Below are the key considerations:

- **Separation from high-traffic areas:**
  - The facility should not be near public spaces and high foot traffic (e.g. reception areas, cafeterias).
  - It should be separated from sensitive areas such as childcare facilities or general medical areas.
- **Accessibility to support systems:**
  - The facility should be located near critical infrastructure such as autoclaves, waste disposal systems, and emergency facilities (*e.g.*, fire suppression systems and eyewash stations).

Fig. 1.1 Biohazard warning notice and sign

- Provide easy access to decontamination facilities such as showers and changing rooms if required.

- **Environmental considerations:**
  - Avoid locating the facility in flood areas prone to flooding or earthquakes.
  - Ensure good ventilation and avoid shared airflow with non-laboratory areas.
- **Proximity to important facilities:**
  - Locate the facility near research laboratories or clinical departments if it is used for infectious disease research.
  - Ensure proximity to utilities such as electricity, water and HVAC (Heating, Ventilation, and Air Conditioning) systems to ensure uninterrupted operation.
- **Restricted access:**
  - Access to the BSL-2 facility should be limited to authorized personnel trained in biosecurity protocols.
  - Use electronic access control systems (e.g., key cards or biometric scanners) to track and restrict access.
- **Signage:**
  - Post clearly visible biohazard signs at entrances indicating the BSL level and specific hazards present.
  - Provide contact information for designated emergency personnel.
- **Visitor policy:**
  - Allow limited visitors and only when accompanied by trained personnel.
  - Maintain an entry and exit log to track visitors.
- **Emergency access:**
  - Ensure that emergency responders have clear instructions on entering the facility during an incident.
  - Provide alternative access routes if the main access points are compromised.
- **Biohazard warning symbol and sign**
  - The biohazard warning sign must be posted on the door(s) of rooms where RG 2 microorganisms are handled, as shown in Fig. 1.1

1.1.2 Surfaces and Floors

- **Seamless and non-absorbent:**

  - All surfaces, including floors, walls, ceilings, countertops, and furniture, must be seamless or have minimal seams to prevent the accumulation of contaminants.
  - Materials should be non-absorbent to prevent the penetration of liquids, chemicals and biological contaminants.
- **Chemical and biological resistance:**
  - Surfaces must withstand frequent contact with disinfectants, alcohols, cleaning agents and biological agents.
  - Flooring should be resistant to corrosion and abrasion from common laboratory chemicals.
- **Ease of cleaning:**
  - Surfaces must allow for quick and thorough cleaning, especially after spills or contact with biohazards.
  - Smooth surfaces are important to reduce cleaning time and ensure effective disinfection.
- **Vinyl or epoxy coatings:**
  - Seamless, chemical resistant, and durable.
  - Epoxy resin is particularly suitable for creating a smooth and impermeable surface.
- **Rubber or polyurethane flooring:**
  - Provides flexibility, slip resistance and easy cleaning.
  - Suitable for areas with heavy equipment and frequent movement.
- **Use of stainless steel:**
  - Commonly used for benches, sinks and countertops because it is corrosion-resistant and easy to clean
- **Laminate with sealed edges:**
  - An inexpensive option, but only if properly sealed to prevent liquid penetration.

1.1.3 Structural Layout

Separate work areas with designated areas for biohazardous material handling, equipment and waste management [1].

Area for Handling Biohazardous Material This area is designated for handling infectious materials and potentially contaminated samples.

Main Features

- **Primary containment equipment:**
  - Biological Safety Cabinets (BSCs), especially Class II are the centrepiece for handling biohazardous materials.

  - These cabinets protect personnel and the environment by preventing aerosolized particles from escaping.
- **Designated work zones:**
  - Clearly labelled areas within the laboratory should be designated exclusively for the handling of hazardous materials only.
  - Use color-coded labels, floor markings, or signs to demarcate these areas.
- **Proximity to essential tools:**
  - Ensure pipettes, microcentrifuges, and other commonly used equipment are located near the BSCs to minimize movement and the risk of accidental exposure.
- **Decontamination:**
  - Place a decontamination station nearby to clean surfaces, tools, and small spills.
  - Provide easily accessible sanitizers such as 70% ethanol, 10% bleach, or other approved agents.
- **Material flow:**
  - **Ensure a unidirectional workflow for samples:** Receipt into the hazardous area, processing in the BSC, and transportation to designated storage or waste areas.

Equipment Work Area

This area houses general lab equipment that does not require a BSC but is still essential for routine lab work.

Main Features

- **Dedicated space for non-hazardous equipment:**
  - Microscopes, spectrophotometers, thermal cyclers, and other non-hazardous equipment should be housed separately.
  - Avoid placing these devices near areas where hazardous material is handled to reduce the risk of contamination.
- **Designated shared equipment:**
  - Clearly label equipment such as centrifuges or incubators as hazardous or non-hazardous.
  - Mark equipment used exclusively for handling infectious materials with biohazard tape or stickers.
- **Cleaning and maintenance:**
  - Establish regular cleaning protocols for equipment in this area to ensure safety and functionality.
  - Place a hand washing station or sanitizer dispenser nearby to encourage frequent hand hygiene
- **Segregation of Use:**
  - Do not allow equipment that handles hazardous materials to be used for other purposes until thoroughly decontaminated.

- **Efficient layout:**
  - Arrange equipment to allow smooth transitions between workflows while maintaining separation from hazardous areas.

Waste Management Area
Proper waste segregation ensures safe disposal and minimizes exposure to hazardous materials.

Key Features

- **Dedicated waste collection zone:**
  - Dedicate a specific area for biohazard waste containers, sharps containers and general waste.
  - Waste containers must be clearly labelled (*e.g.*, biohazard symbols for infectious waste).
- **Access to the autoclave:**
  - Ideally, the waste disposal area should be adjacent to or have direct access to an autoclave
  - This facilitates the immediate decontamination of biohazardous waste.
- **Disposal of liquid waste:**
  - Provide designated sinks for liquid waste disposal, equipped with chemical sanitizers for pretreatment.
- **Waste separation:**
  - Use separate containers for the different types of waste: biological, chemical, and general waste.
  - Ensure containers are color-coded and sealed to prevent accidental spills.
- **Transportation routes:**
  - Create clear and safe routes for waste transportation from work areas to the waste disposal zone.
  - Avoid crossing clean areas during transportation.

1.2 Personal Protective Equipment (PPE)

PPE is an important part of safety protocols in Biosafety Level 2 (BSL-2) laboratories. Proper use of PPE minimizes the risk of exposure to infectious agents and protects laboratory personnel from accidental contamination. Standard PPE for BSL2 is discussed in the following section:

1.2.1 Lab Coats

Lab coats are an essential component of PPE in BSL-2 facilities. They form a protective barrier between the worker and biohazardous materials, reducing the risk of exposure to infectious agents and accidental contamination. The key requirements for lab coats in BSL-2 are detailed below in detail:

Material

- **Fluid resistant fabric:**
  - Lab coats should be made of polyester-cotton blends with liquid-resistant coatings to protect against splashes of biohazardous liquids.
- **Durability:**
  - Choose materials that can withstand frequent laundering and autoclaving without significant wear or deterioration.

Design

- **Long sleeves:**
  - Coats must have long sleeves with elasticized or knitted cuffs to protect the skin from exposure.
- **Full coverage:**
  - The coat should reach at least mid-thigh to provide adequate coverage.
- **Type of closure:**
  - Front closures should be fastened with snaps or ties as they are easier to remove in case of contamination. Avoid coats with zippers for this reason.
- **Pockets:**
  - Pockets are acceptable but should be kept to a minimum or avoided to prevent them from harboring contaminants.

Fit

- **Comfortable fit:**
  - Ensure coats are not too tight so as not to restrict movement or too loose so as not to catch on equipment
- **Sizing:**
  - A variety of sizes should be available to accommodate all personnel/workers.

Color Coding (Optional)

- Consider using different colors for specific purposes or areas (*e.g.*, white for general work, blue for hazardous areas) to visually indicate the level of contamination risk.

Putting on and Taking off Clothing

- **Before entering:**
  - Put on a lab coat before entering the laboratory work area.
- **After use:**
  - Remove the lab coat before leaving the laboratory and dispose of or store it properly to avoid contaminating other areas.

Storage

- **Separate clean and used lab coats:**
  - Store clean lab coats in a designated area and use lab coats in biohazard laundry containers.

Decontamination

- **Frequent cleaning:**
  - Lab coats should be laundered regularly (*e.g.*, weekly or when visibly soiled) in a facility equipped to handle biohazardous materials.
- **Autoclaving:**
  - Coats contaminated with infectious materials should be autoclaved before laundering.

Replacement

- Replace coats that are damaged, torn or no longer fit properly to ensure that protection is not compromised

1.2.2 Gloves

Gloves are an important part of PPE in BSL-2 facilities. They protect laboratory personnel from direct contact with hazardous materials, chemicals, and contaminated surfaces, significantly reducing the risk of exposure and contamination. The key considerations for glove use are as follows:

Material

- **Nitrile gloves:**
  - Preferred for their resistance to punctures, chemicals, and infectious agents.
  - Provide excellent sensitivity and dexterity for detailed work.

- **Latex gloves:**
  - Provide good flexibility and tactile sensitivity but may cause allergic reactions in some individuals.
  - Non-powdered latex gloves are recommended to avoid contamination.
- **Vinyl gloves:**
  - Suitable for short-term, low-risk work, but less durable and less resistant to chemicals and punctures.

Thickness

- Gloves are typically between 4 to 8 mils thick:
  - Thinner gloves (4–6 mils) for jobs requiring dexterity.
  - Thicker gloves (6–8 mils) for handling sharp objects or potentially hazardous chemicals.

Properties

- **Disposable and single-use:**
  - Gloves should be disposable to avoid cross-contamination.
  - Reuse of gloves is not permitted in BSL-2 facilities.
- **Powder-free:**
  - Powder-free gloves minimize airborne contamination and prevent skin irritation.
- **Ambidextrous or hand specific:**
  - Ambidextrous gloves for general use; hand-specific gloves for tasks requiring precision.

Donning Gloves

- Check gloves for tears, holes or defects before use.
- Make sure the gloves cover the cuffs of the lab coat to ensure seamless protection.

Use of Gloves

- Always wear gloves when handling the following materials:
  - Biohazardous materials.
  - Contaminated surfaces or equipment.
  - Chemicals or reagents.
- Avoid touching non-laboratory objects (*e.g.*, phones, door handles) to prevent contamination.

Removal of Gloves

- Remove gloves carefully to avoid touching the contaminated surface.
- Dispose of used gloves immediately in biohazardous waste containers.

Storage

- Store gloves in a clean, dry place away from direct sunlight, heat, and chemicals that could degrade the material.

Disposal

- Dispose of used gloves in the designated bio-waste containers.
- Do not dispose of gloves in normal waste if they are contaminated with hazardous materials.

1.2.3 Eye and Face Protection

In BSL-2 laboratories, eye and face protection is essential when working with infectious substances, chemicals or processes that generate aerosols or splashes. Appropriate eye and face protection minimizes the risk of exposure to hazardous materials and contaminants.

Safety Goggles

- **Features:**
  - Provides a tight seal around the eyes for complete protection.
  - Prevents exposure to aerosols, splashes and fine particles.
  - Anti-fog coating for better vision during prolonged use.
- **Use:**
  - Recommended when working with liquid biohazards, splashes or processes that generate aerosols.

Face Masks

- **Features:**
  - Complete coverage of the face from the forehead to under the chin.
  - Protects against large amounts of splashes and flying debris.
  - Must be used with goggles when protection against fine aerosols is required.

- **Usage:**
  - Ideal for high-risk work involving large quantities of hazardous liquids or heavy aerosol generation.

Safety Goggles

- **Features:**
  - Provides basic eye protection against direct splashes and particulates.
  - Lightweight and comfortable for general use.
  - Can be fitted with side shields for enhanced side protection.
- **Uses:**
  - Suitable for low-risk work but not recommended when working with highly infectious substances or aerosols.

Respiratory Protection

- Masks, such as N95 respirators, may be required when working with agents that generate aerosols.
- Tested respirators for work with a higher risk of inhalation exposure.

Protection for Head and Feet

- Hair protection to prevent hair contamination, especially for sensitive experiments.
- Closed, non-slip shoes to protect against spilled liquids or sharp objects. Shoe covers may be used as an additional protective measure

1.3 Decontamination Procedures

Decontamination is a critical aspect of BSL-2 laboratory operations to ensure that infectious agents are rendered harmless and contamination of personnel, equipment and the environment is prevented [2]. Some key decontamination procedures are discussed below:

Surface Decontamination

- Work surfaces should be disinfected before and after experiments, and immediately after spills or splashes.
- Use appropriate disinfectants, such as 10% sodium hypochlorite, 70% ethanol/isopropanol or commercially available disinfectants that are effective against the substances to be treated.
- Ensure sufficient exposure time to the disinfectant manufacturer's instructions.

Decontamination of the Equipment

- **Biosafety cabinets (BSCs):** Clean and decontaminate internal surfaces regularly with disinfectants.
- Instruments: Autoclave or disinfect reusable instruments after each use.
- Centrifuges: Decontaminate rotors and buckets if they have encountered biological substances.

Decontamination of Waste

- **Biological waste**: Autoclave solid waste such as culture plates, contaminated gloves and pipette tips before disposal.
- **Liquid waste**: Treat with disinfectants such as bleach (final concentration 1:10) for at least 20 min before disposal into the sewer (if local regulations permit).

Spill Response

- Isolate the spill area immediately and notify personnel.
- Cover the spill with absorbent material (*e.g.*, paper towels) and soak with a disinfectant, allowing sufficient contact time.
- Dispose of the waste as hazardous material.

Personal Decontamination

- Wash your hands thoroughly with soap and water after handling biological agents and before leaving the laboratory.
- If it has come in contact with your skin, rinse it immediately with water and clean it with soap or an antiseptic.

Decontamination of the Facility

- Routine cleaning of high-contact areas such as door handles, light switches, and work surfaces.
- Regular deep cleaning and fumigation (*e.g.*, with hydrogen peroxide vapor or formaldehyde) depending on laboratory use and risks.

1.4 Health and Medical Surveillance

Health and medical surveillance programs are essential for BSL-2 laboratory operations. These programs aim to protect laboratory personnel from occupational exposure to biological hazards and to provide immediate medical care in the event of accidental exposure. Some important aspects are the following:

Pre-employment Screening

- Basic health assessment to determine suitability for working with biohazards.
- Review of medical history emphasizing allergies, immunosuppression or chronic diseases.
- Records of immunizations and their boosters (*e.g.*, hepatitis B vaccination).

Immunization Programs

- Vaccinations against pathogens frequently handled in the BSL-2 laboratory, e.g. hepatitis B virus or influenza virus.
- Post-vaccination antibody titer testing to confirm immunity, if required.

Training and Awareness

- Regular training on biosecurity practices, emergency protocols, and exposure management.
- Awareness programs to recognize the symptoms of laboratory-acquired infections.

Occupational Exposure Management

- Immediate reporting and assessing incidents such as needlestick injuries, spills, or aerosol exposures.
- Post-exposure prophylaxis (PEP) for certain pathogens as healthcare professionals recommend.

Regular Health Monitoring

- Routine health checks for early detection of potential occupational diseases.
- Additional monitoring for personnel working with high-risk agents or techniques that generate aerosols.

Incident Response and Record-Keeping

- Keep detailed records of exposure incidents, medical assessments and treatments.
- Confidential medical records must be kept secure and only accessible to authorized personnel.

Medical Assessment on Exit

- Evaluation of personnel's health status upon leaving employment or laboratory duties.

1.5 Risk Assessment and Management

Risk assessment and management are essential processes in BSL-2 laboratories to identify, assess and mitigate potential hazards associated with the handling of medium-risk biological agents. These procedures ensure personnel safety, the environment and the public while complying with biosafety regulations. The various steps of risk assessment and risk management are as follows:

1.5.1 Steps in Risk Assessment

Identify the Biological Agents

- Determine the pathogens you handle, such as Salmonella spp., hepatitis B virus or influenza virus.
- Assess their infectivity, pathogenicity, transmission routes, and environmental stability.

Evaluation of Laboratory Activities

- Investigate procedures that generate aerosols, use sharps or generate large quantities of cultures.
- Evaluate the likelihood and severity of exposure in routine and accident

Identify Personnel at Risk

- Evaluate health conditions (*e.g.* immunosuppression, pregnancy) that may increase susceptibility.
- Evaluate the training and competence of laboratory staff.

Review Facility and Equipment

- Ensure proper design, containment features, and maintenance of BSCs, autoclaves, and ventilation systems.

Analyze Environmental Risks

- Consider the potential for accidental release and its impact on the environment.

1.5.2 Steps in Risk Management

Engineering Controls

- Use certified BSCs (Class II) for infectious agent procedures.
- Ensure directional airflow and HEPA filtration for ventilation systems.
- Provide autoclaves for waste decontamination.

Administrative Controls

- Develop and enforce standard operating procedures (SOPs).
- Restrict laboratory access to authorized and trained personnel.
- Provide regular biosafety training and competency assessments.

PPE

- Require the use of lab coats, gloves, eye protection, and masks as appropriate.
- Ensure proper donning and doffing practices to prevent contamination.

Emergency Preparedness

- Establish protocols for spills, accidental exposures, and evacuations.
- Provide immediate access to first aid, eyewash stations, and safety showers.

Health Surveillance

- Conduct pre-employment health assessments and maintain immunization programs.
- Monitor and respond to potential laboratory-acquired infections.

1.6 Documentation and Record Keeping

Comprehensive documentation and record-keeping are critical components of BSL-2 laboratory operations. These practices ensure compliance with biosafety regulations, facilitate effective management of laboratory activities, and provide a reliable framework for incident response and audits. Key areas of documentation and record keeping are discussed below:

Biosafety Protocols and Procedures

- Maintain SOPs for handling biological agents, equipment uses, and emergency responses.
- Update SOPs periodically to reflect changes in procedures, regulations, or laboratory practices.

Training Records

- Document training programs attended by laboratory personnel, including biosafety, PPE use, and emergency protocols.
- Include dates, training content, and trainer certifications to verify compliance and competency.

Risk Assessments

- Record risk assessments for biological agents, equipment, and procedures.
- Include hazard identification, exposure likelihood, and mitigation measures.

Health and Medical Surveillance

- Maintain confidential health records for personnel, including pre-employment screenings, immunizations, and post-exposure evaluations.
- Document incidents of laboratory-acquired infections or exposures.

Agent and Material Inventory

- Keep detailed logs of biological agents, chemicals, and other hazardous materials stored and used in the laboratory.
- Include information on quantities, storage conditions, and disposal methods.

Decontamination Records

- Document routine cleaning and decontamination of surfaces, equipment, and waste.
- Record autoclave cycles, including temperature, pressure, and duration, to verify efficacy.

Incident and Accident Reports

- Maintain detailed reports of spills, exposures, or accidents, including the cause, response actions, and preventive measures implemented.

Equipment Maintenance and Calibration

- Record routine maintenance and calibration of BSCs, autoclaves, and other critical equipment.
- Include service dates, technician details, and calibration outcomes.

Waste Management Logs

- Document the collection, treatment, and disposal of biohazardous waste.
- Ensure compliance with local and institutional waste disposal regulations.

Audits and Inspections

- Record findings from internal and external audits or inspections.
- Document corrective actions taken to address identified deficiencies.

1.7 Training and Education

Training and education are fundamental to ensuring the safe and efficient operation of BSL-2 laboratories. These programs equip personnel with the knowledge and skills to handle biohazards responsibly, comply with safety protocols, and respond effectively to emergencies.

1.7.1 Objectives of Training and Education

The objectives of training and education for BSL-2 guidelines are designed to ensure the safe handling of potentially infectious materials, compliance with regulatory standards, and the development of a culture of biosafety within the laboratory environment. Key objectives include:

Enhance Biosafety Knowledge

- Provide an understanding of BSL-2 hazards, including the types of pathogens and risks associated with working in such environments.
- Familiarize personnel with relevant regulations, institutional policies, and national/international biosafety standards.

Develop Safe Laboratory Practices

- Train personnel in SOPs for handling biological agents.
- Emphasize the proper use of PPE to minimize exposure risks.
- Promote best practices for decontamination, waste management, and spill response.

Build Competence in Emergency Preparedness

- Prepare personnel to respond effectively to:
  - Accidental exposures
  - Spills
  - Injuries
  - Other lab incidents
- Train on appropriate reporting and follow-up protocols in case of emergencies.

Ensure Safe Equipment Handling

- Educate on the correct use and maintenance of biological safety cabinets, centrifuges and other lab equipment.
- Ensure understanding of safety measures to prevent aerosols, leaks and equipment malfunctions.

Promote Compliance with Biosafety Guidelines

- Foster adherence to IBSC rules and recommendations.
- Ensure all activities align with the Occupational Safety and Health Administration (OSHA) and Centers for CDC standards.

Cultivate a Culture of Biosafety

- Encourage the adoption of a safety-first mindset among all lab personnel.
- Promote teamwork, accountability, and a commitment to biosafety as part of routine laboratory operations.

Address Specific Risks

- Provide targeted training for:
  - Handling sharps safely.
  - Transporting infectious materials securely.
  - Identifying and mitigating risks unique to the pathogens or procedures used in the lab.

Continuous Learning and Assessment

- Enable personnel to stay updated on new pathogens, techniques, and biosafety challenges.
- Conduct regular assessments, including quizzes, practical drills, and audits, to ensure training effectiveness and compliance.

1.7.2 Key Components of BSL-2 Training Programs

- **Introduction to biosafety**
  - Overview of BSL-2 laboratory design and containment features.
  - Explanation of the risk group classification of biological agents.
- **SOPs**
  - Step-by-step guidance on laboratory practices, including proper use of BSCs, handling and disposal of biohazardous materials and decontamination and waste management.
- **PPE**
  - Instructions on the selection, proper use, and disposal of PPE.
  - Hands-on practice in donning and doffing PPE to prevent contamination.
- **Risk assessment**
  - Training on identifying hazards, assessing risks, and implementing appropriate controls.
  - Case studies on laboratory incidents to reinforce learning.
  - **Emergency response:** It should include procedures for managing spills, exposures, and evacuations, including eyewash stations, safety showers, and first-aid kits.
- **Health and safety protocols**
  - Importance of immunizations and medical surveillance programs.

  - Personal hygiene practices, including hand washing techniques.
- **Documentation and record keeping**
  - Importance of maintaining accurate records of training, incidents, and laboratory activities.
  - Use of digital tools for efficient documentation.
- **Periodic refresher courses**
  - Updates on new biosafety guidelines, equipment, and techniques.
  - Reassessment of competencies to ensure skills remain current.

1.7.3 Methods of Delivery

Delivering training and knowledge of BSL-2 guidelines requires methods that are effective, engaging, and tailored to the learners' needs. A mix of approaches ensures a comprehensive understanding of biosafety principles and practices. Below are the key methods:

Classroom-Based Training

- **Lectures and presentations:**
  - Provide structured explanations of BSL-2 guidelines, including hazards, safety protocols, and emergency responses.
  - Use visuals like slides, videos, and infographics to enhance understanding.
- **Interactive discussions:**
  - Encourage dialogue and Q&A sessions to clarify doubts and explore real-world scenarios.

Hands-on Practical Training

- **Laboratory simulations:**
  - Conduct mock exercises to demonstrate safe handling of biological agents, use of PPE, and decontamination techniques.
- **Equipment demonstrations:**
  - Show proper usage and maintenance of BSCs, centrifuges, and autoclaves.
- **Emergency drills:**
  - Practice spill response, exposure management, and evacuation procedures.

Online and e-Learning Modules

- **Virtual courses:**
  - Offer self-paced modules covering key biosafety topics.
- **Interactive content:**
  - Include quizzes, videos, and animations to keep learners engaged.

- **Webinars:**
  - Host live online sessions for updates, Q&A, and expert talks on biosafety.

Written Materials and Guidelines

- **Manuals and SOPs:**
  - Provide detailed written resources, including SOPs and safety guidelines.
- **Quick reference cards:**
  - Distribute concise materials for quick reminders of key biosafety practices.

Case Studies and Problem-Solving Sessions

- **Real-life examples:**
  - Analyze case studies of biosafety breaches to discuss lessons learned.
- **Scenarios and role-playing:**
  - Engage learners in identifying risks and proposing solutions in hypothetical situations.

On-the-Job Training

- **Mentorship programs:**
  - Pair new personnel with experienced staff for hands-on guidance.
- **Shadowing:**
  - Allow trainees to observe experienced professionals in action.
- **Incremental learning:**
  - Gradually introduce new responsibilities as competence grows.

Assessments and Feedback

- **Knowledge tests:**
  - Use written quizzes or online tests to evaluate understanding.
- **Practical evaluations:**
  - Assess skills through observed performance during drills and simulations.
- **Feedback sessions:**
  - Provide constructive feedback to address gaps in knowledge or practices.

Interactive Tools

- **Game-based learning:**
  - Use games or simulations to make training engaging while reinforcing concepts.

- **Virtual reality (VR):**
  - Offer immersive experiences for practicing biosafety in a risk-free environment.

Regular Refresher Courses

- **Periodic updates:**
  - Reiterate key practices and incorporate new regulations or emerging risks.
- **Continuing education:**
  - Provide advanced training for experienced personnel to stay updated.

Peer Reviews and Group Discussions

- Encourage collaborative learning by discussing incidents, sharing best practices, and fostering a culture of safety awareness.

1.7.4 Evaluation and Certification

Evaluation and certification are essential for BSL-2 training programs. They ensure that personnel understand and can apply biosafety principles effectively. Below are the methods for evaluating and certifying BSL-2 training and knowledge:

Evaluation Methods

Knowledge Assessment

- **Written tests:** Multiple-choice questions, true/false, and short-answer questions covering:
  - Identification of risks.
  - Biosafety protocols.
  - Emergency response procedures.
- **Online quizzes:**
  - Automated testing platforms for immediate feedback.

Practical Skills Assessment

- **Hands-on demonstrations:** Trainees demonstrate correct techniques, such as:
  - Using PPE properly.
  - Operating BSCs.
  - Managing spills and waste.
- **Simulated scenarios:** Evaluate responses to mock emergencies, such as:

 - Exposure to infectious agents.
 - Equipment malfunctions.

Behavioural Observation

- **Performance monitoring:**
  - Supervisors observe compliance with safety protocols during daily lab work.
- **Peer reviews:**
  - Colleagues provide feedback on adherence to best practices.
- **Case study analysis**
  - Trainees analyze biosafety incidents and suggest improvements, assessing their understanding of risk management and problem-solving skills.
- **Feedback sessions**
  - Post-training interviews or discussions to gauge comprehension and address gaps in knowledge.

Certification Process

Completion Criteria

- **Minimum passing scores:**
  - Set thresholds for written exams and practical assessments.
- **Participation:**
  - Require attendance in all training modules, workshops, and simulations.

Issuing Certificates

- Certificates should include:
  - Trainee's name.
  - Course title and content summary.
  - Training completion date.
  - Certification validity period (*e.g.*, 1 or 2 years).
  - Signature of the training authority or biosafety officer.

Certification Renewal

- **Refresher courses:**
  - Mandate periodic retraining to maintain certification and incorporate updates in biosafety standards.
- **Continuous evaluation:**
  - Require passing updated assessments to renew certification.

Importance of Certification

- **Compliance:** Meets regulatory requirements (*e.g.*, CDC, OSHA).
- **Accountability:** Ensures personnel understand and accept their responsibility for biosafety.
- **Recordkeeping:** Certification records provide proof of training for audits and inspections.
- **Competency verification:** Confirms that staff can safely work with BSL-2 materials and equipment.

Follow-up Measures

- **Post-certification monitoring:** Regular performance evaluations to ensure continued adherence to BSL-2 practices.
- **Incident Reporting:** Use real-world feedback from incidents or near-misses to improve training programs.

By combining thorough evaluations with recognized certification processes, institutions can ensure that personnel meet the high standards required for working in BSL-2 environments safely and effectively.

Therefore, BSL-2 guidelines are essential for safely handling moderate-risk biological agents in laboratory environments. These guidelines protect laboratory personnel, the environment, and the community from potential exposure to pathogens while enabling productive scientific research. The implementation of BSL-2 practices is based on a well-established framework that includes facility design, SOPs, risk assessment, PPE, health surveillance, and ongoing education and training. The BSL-2 guidelines emphasize a multi-layered approach to safety, starting with the design of the laboratory itself. The structural integrity of BSL-2 labs is designed to contain biological agents and reduce exposure risks. Features such as controlled access, proper ventilation with directional airflow, BSCs and effective waste disposal systems play a critical role in containment. These physical barriers are complemented by operational procedures and protocols that guide researchers in working safely with pathogens. Proper risk assessment and management ensure that all potential hazards are identified and mitigated; reducing the likelihood of accidents and exposure incidents. A core element of BSL-2 safety is the use of PPE. PPE is the first defense against contamination, preventing direct contact with biological agents. The appropriate use of lab coats, gloves, safety goggles, face shields, and, in some cases, respiratory protection ensures that personnel remain protected while handling infectious materials. Furthermore, regular decontamination procedures and proper waste management protocols are essential to ensure the workspace and waste products are properly sanitized and disposed of, minimizing the risk of cross-contamination and environmental release.

Health and medical surveillance are another integral aspect of BSL-2 laboratory guidelines. The early detection of potential health issues resulting from occupational exposure is vital for protecting laboratory personnel. Pre-employment health screenings, immunization programs, and continuous health monitoring ensure that employees are fit for work and adequately protected against the risks of handling biological agents. Furthermore, in the event of an exposure incident, a robust system for reporting and responding to accidents ensures that any health concerns are promptly addressed. Equally important is the continuous education and training of laboratory staff. Comprehensive training programs cover a range of topics, including biosafety practices, risk management, emergency protocols, and the use of PPE. Ongoing education helps reinforce safe practices, ensures staff members remain aware of new hazards or regulations, and maintains high competency standards within the laboratory. By keeping personnel well-informed and well-prepared, training minimizes the risk of human error and fosters a culture of safety within the laboratory.

Thus, the BSL-2 guidelines are a comprehensive and structured set of protocols that provide a framework for safely handling biological agents. Adherence to these guidelines is essential to minimizing risks to personnel, the environment, and the public while ensuring the continuation of valuable research. The success of BSL-2 laboratories relies on the collective responsibility of laboratory management, staff, and regulatory bodies to implement and maintain safety measures effectively. By combining strong physical controls, thorough training, ongoing risk assessments, and a culture of continuous improvement, BSL-2 laboratories can provide a safe and productive environment for scientific discovery.

References

1. Guidelines on handling, transportation, and disposal of biomedical waste, critical for BSL-2 operations. Government of India, 2024
2. National guidelines for biological disaster management. NDMA, Government of India, 2024. Highlights containment strategy and biosafety protocols in managing infectious agents

Handling, Culturing, Preservation and Revival of Fungal Pathogens

2

Abstract

Fungal pathogens pose a major challenge to agricultural and biological research and require effective methods for their handling, cultivation, preservation and resuscitation. The handling of fungal pathogens requires strict adherence to biosafety protocols to prevent contamination and ensure laboratory safety. Proper techniques, such as aseptic transfer and the use of laminar flow cabinets, are essential for maintaining pure cultures. Cultivation of fungal pathogens requires the selection of appropriate growth media, which vary depending on the nutritional and environmental requirements of the pathogen (Pitt and Hocking, Fungi and food spoilage, 3rd edn. 2009). Common media include potato dextrose agar (PDA) and yeast extract peptone dextrose (YEPD) supplemented with specific nutrients or inhibitors to promote fungal growth while suppressing contaminants. Environmental conditions such as temperature, humidity and pH are critical for optimizing fungal development. Preservation methods are critical for maintaining the viability and genetic stability of fungi over time. Techniques such as cryopreservation at −80 °C and freeze-drying (lyophilization) are often used for long-term preservation. For short-term storage, fungal cultures can be stored on agar plates or in sterile distilled water at suitable temperatures (Nakasone et al, Biodiversity of fungi: inventory and monitoring methods 2004). These methods help to preserve the pathogenic and phenotypic characteristics required for subsequent studies. To revive preserved fungal cultures, they must be carefully thawed and cultured under optimal conditions. This process includes rehydration, inoculation on fresh culture media and incubation in a growth-promoting environment. Particular attention is paid to recovering viable and uncontaminated cultures, which is crucial for subsequent experiments. Molecular and morphological studies can be performed to confirm the identity and integrity of the recovered pathogen. Effective management of fungal pathogens through these techniques supports advances in disease diagnostics, pathogen resistance

R. Pasrija, D. Kumari, *Protocols for Studying Pathogenic Fungi*,
https://doi.org/10.1007/978-981-96-7047-5_2

research and the development of sustainable agricultural practices. Continued refinement of these techniques, along with the integration of molecular tools, improves our understanding and control of fungal pathogens in diverse systems.

Keywords

Fungi · Pathogen · Handling · Cultivation · Conservation · Revitalization · Protocols

Introduction In this chapter, we shall be dealing with some basic techniques for handling, cultivating, preserving and reviving fungal pathogens. We will cover the following topics as follows:

2.1. Handling Cultures of Fungal Pathogens
2.2. Cultivation of Fungal Pathogens
2.3. Preservation of Fungal Cells
2.4. Resuscitation of Fungal Cells

2.1 Handling Cultures of Fungal Pathogens

Properly handling fungal pathogens requires careful attention from the initial phase of isolation or revival of fungi to the final phase of preservation. Failure to follow correct procedures significantly increases the risk of contamination of pure fungal cultures. It is important to maintain general hygiene and take preventative measures, starting with inspecting all materials entering the laboratory. The most important guidelines for handling fungal cultures are listed below:

2.1.1 Biosafety in the Laboratory and Personal Protective Measures

Fungal pathogens, especially those with zoonotic potential or high virulence, require strict biosafety protocols. Personnel must wear appropriate personal protective equipment (PPE) such as lab coats, gloves and masks to minimize the risk of exposure. Working in a biosafety cabinet (laminar flow hood) is essential for handling fungal cultures as it prevents the spread of aerosols and reduces the risk of contamination. Proper sterilization of work surfaces and tools with disinfectants or by autoclaving ensures a sterile working environment.

2.1.2 Aseptic Techniques

Maintaining a sterile work area is critical when working with fungal cultures. Aseptic techniques, such as using sterilized tools (snares, scalpels and pipettes) and performing all manipulations within a sterile zone, prevent cross-contamination. The use of flame or alcohol to sterilize the instruments during transfer further ensures the purity of the fungal cultures.

2.1.3 Selection and Preparation of Culture Media

Different fungal pathogens have specific nutritional and environmental requirements. The selection of a suitable growth medium is essential for successful cultivation. Common media include:

- **Potato Dextrose Agar (PDA)/Potato Dextrose Broth (PDB):** A general-purpose medium suitable for a wide range of fungi.
  - **Composition and preparation:** To make 1 L of PDA, dissolve 200 g potato infusion, 20 g dextrose and 15 g agar, or alternatively, mix 39 g commercial PDA powder with distilled water to 1 L. Autoclave at 121 °C and 15 psi for 15 min. For PDB preparation do not add agar in the remaining constituents or mic 24 g commercial PDB with distilled water to 1 L. Adjust the pH to 5.6 (±0.2), if necessary, using HCl or NaOH.
- **Yeast Extract Peptone Dextrose (YPD or YEPD):** The "Y" or "YE" in YEPD refers to a yeast extract that contains the water-soluble compounds produced when yeast is forced to self-digest. "P" refers to peptone, a mixture of peptides and amino acids produced by digesting animal protein with proteases. "D" refers to dextrose, or glucose, a carbon source for yeast. YPD contains essential nutrients to fulfill the metabolic requirements of cells. It is a general medium that promotes yeast growth.
  - **Composition and preparation of YEPD:** To prepare 1 L of YEPD, mix 10 g of yeast extract, 20 g of peptone, 20 g of dextrose and 15 g of agar with distilled water to 1 L. Autoclave at 121 °C and 15 psi for 15 min.
- **Synthetic Drop-out Medium:** It is used for screening various nutrient-deficient fungi, selecting yeast transformants, yeast-2-hybrid screening, etc. Various drop-out (DO) supplements are added to the Minimal synthetic defined (SD) bases to produce selective media for yeast culture growth, confirmation of auxotrophic phenotypes and selection of yeast transformants. These supplements could be uracil, leucine, tryptophan, etc.
  - **Composition and preparation of Synthetic Drop-out medium:** Mix 6.7 g of nitrogen base without amino acids (0.67%), 0.87 g of amino acid drop-out mix, 20 g of dextrose (2%) and 20 g of agar with distilled water to 1 L. Autoclave at 121 °C and 15 psi for 15 min.
    - **Amino acid drop-out mix for SD:**

Adenine	800 mg
Uracil	800 mg
Tryptophan	800 mg
Histidine	800 mg
Arginine	800 mg
Methionine	800 mg
Tyrosine	800 mg
Leucine	800 mg
Lysine	800 mg
Phenylalanine	800 mg
Threonine	800 mg

If needed (only add if the strains contain these markers):

Isoleucine	1200 mg
Valine	6000 mg
Aspartatic acid	4000 mg

- **RPMI 1640 Agar with MOPS (3-morpholino propane sulfonic acid) & 2% Glucose without Sodium bicarbonate:** It is used to determine the sensitivity of microorganisms to antifungal agents. RPMI-1640 medium was developed by Moore et al. [2] at Roswell Park Memorial Institute and is a well-known medium. Invasive fungal infections have increased dramatically in recent times. Due to the life-threatening nature of these infections and reports of drug resistance, susceptibility testing of yeast pathogens has become very important. The CLSI (Clinical & Laboratory Standards Institute) has published a reference method for broth dilution antifungal susceptibility testing in broth dilution for yeasts.
  - **Composition and preparation of RPMI 1640 Agar w/MOPS & 2% glucose without sodium bicarbonate**: Dissolve 10.4 g RPMI-1640 powder (with glutamine and phenol red, but without bicarbonate) in 900 mL distilled water. Add 34.53 g MOPS (0.165 mol/L final) and adjust the pH to 7.0 with NaOH. Makeup to a final volume of 1 L with distilled water. Filter sterilizes the media and stores it at 4 °C.
- **Selective media:** Often enriched with antibiotics or fungicides to isolate specific fungi.
- **Note:** Media must be prepared under sterile conditions, paying attention to pH, nutrient composition and sterilization procedures to ensure fungal growth without contamination

2.1.4 Culture Maintenance and Subcultures

Fungal cultures should be subcultured regularly to maintain viability and avoid nutrient depletion of the medium. During subculture, mycelial fragments or spores are transferred to fresh media under sterile conditions. The timing and frequency of the transfer depend on the growth rate of the fungus and the intended experimental use.

2.1.5 Storage and Transportation of Cultures

For short-term storage, cultures can be kept on agar plates at 4–10 °C. For longer periods, more robust preservation techniques, such as cryopreservation or lyophilization, are recommended. When transporting fungal cultures, secure packaging in leak-proof containers is required to prevent accidental release and contamination during transportation.

2.1.6 Disposal of Biosafety Waste

Fungal cultures should be disposed of properly after handling to avoid contamination of the environment. Contaminated materials, including media and tools, must be autoclaved before disposal. Biosafety guidelines require proper labeling and containment of waste materials.

2.1.7 Documentation and Records

Accurate records of culture origin, handling protocols and experimental results are critical for reproducibility and traceability. Proper labeling of the cultures with strain information, date and media type is essential to avoid confusion.

Handling fungal cultures, therefore, requires meticulous attention to detail and strict adherence to aseptic and biosafety protocols. By following best practices, researchers can ensure the purity of fungal cultures, minimize contamination risks and ensure the safety of laboratory personnel. These efforts not only facilitate accurate research results but also contribute to the broader goal of controlling fungal pathogens in agricultural and clinical settings.

2.1.8 Troubleshooting

1. **Problem: Contamination in cultures**
   - **Possible causes**
     - Inadequate sterilization of media, tools or work area.
     - Contaminated fungal strains or inoculum.
     - Inappropriate aseptic techniques.
   - **Solutions**
     - Autoclave all media, glassware and tools thoroughly before use.
     - Work in a biological safety cabinet and clean surfaces with 70% ethanol or bleach before starting.

- Use sterile pipettes and tips, and ensure that the fungal inoculum is derived from a pure culture.
- Avoid exposing the plates or vials to air for long periods of time.

2. **Problem: Difficulties in preparing the spore suspension:** Some fungi, including *Aspergillus*, *Rhizopus*, *Fusarium*, *etc.*, are known as molds and make spores.
   - **Possible causes**
     - Improper scraping techniques for spores.
     - Insufficient mixing or clumping of spores.
   - **Solutions**
     - Use a sterile loop or cell scraper to gently dislodge spores from agar surfaces.
     - Add a small volume of sterile saline or Tween-20 (0.01–0.1%) to help disperse spores.
     - Pass the suspension through sterile gauze or a nylon filter to remove clumps.
     - For example, the cultures and spores of different fungi are shown in Fig. 2.1.
3. **Problem: Poor or delayed growth of fungi**
   - **Possible causes**
     - Incorrect growth medium or conditions.
     - Low viability of fungal spores or cultures.
     - Contamination with antifungal compounds or residual disinfectants.
   - **Solutions**
     - Use a growth medium that is optimized for the respective fungus (*e.g.* YEPD for *Candida* sp.).
     - Check the viability of the fungal strain with a fresh strain or culture.
     - Ensure that tools and surfaces are free of disinfectant residues by rinsing them with sterile water after cleaning.

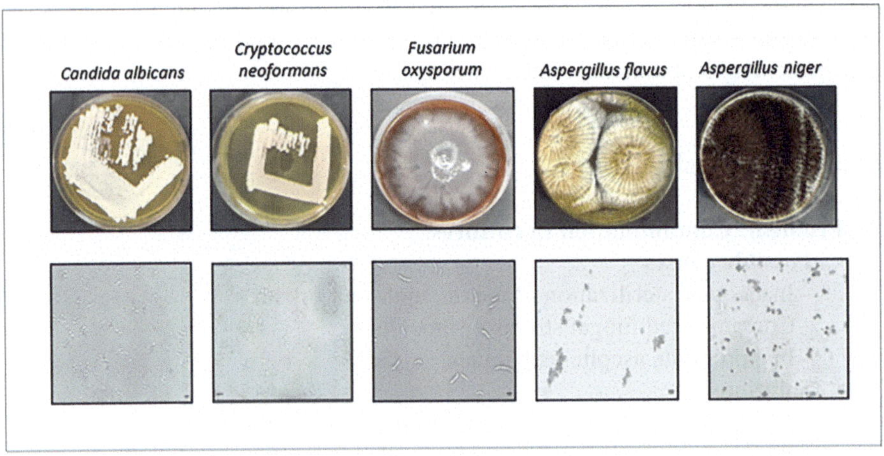

Fig. 2.1 Cultures and spores (40× under light microscope) of different yeasts and molds

4. **Problem: Problems with sporulation**
   - **Possible causes**
     - Incorrect incubation conditions (*e.g.* temperature, humidity or exposure to light).
     - Imbalance of nutrients in the medium. Auxotrophic cultures or some mutants need to be supplemented for growth. e.g. uracil, leucine, arginine
   - **Solutions**
     - Incubate cultures under conditions conducive to sporulation (*e.g.*, 25–30 °C, light-dark cycles if required for certain fungi).
     - Use sporulation-inducing media such as PDA.
5. **Problem: Overgrowth or loss of distinct colony morphology**
   - **Possible causes**
     - Incubated too long or at the wrong temperature.
     - High concentration of initial inoculum.
   - **Solutions**
     - Monitor the cultures regularly and stop incubation when distinct colonies appear.
     - Use calibrated inoculums to avoid overloading the medium with spores or cells.
6. **Problem: Aerosol formation during handling**
   - **Possible causes**
     - Improper handling during pipetting, centrifugation or vortexing.
     - Use of dry fungal spores without proper precautions.
   - **Solutions**
     - Work in a Class II biological safety cabinet when working with cultures that can form aerosols.
     - Use sealed centrifuge tubes and safety rotors.
   - Lightly moisten the spores to reduce aerosol formation and handle them carefully.
7. **Problem: Resistance to antifungal treatments**
   - **Possible causes**
     - Incorrect identification of the fungal strain.
     - Selection pressure leads to resistant subpopulations.
   - **Solutions**
     - Confirm the identity of the fungus using molecular or phenotypic methods.
     - Test antifungal susceptibility using standardized protocols (*e.g.* CLSI or EUCAST methods).
     - Change antifungal agents to minimize the development of resistance.
8. **Problem: Evaporation or desiccation of cultures**
   - **Possible causes**
     - Incubation of plates in a dry environment.
     - Loose or improperly sealed culture vessels.
   - **Solutions**
     - Incubate plates in a humidified chamber or use moist paper towels in sealed containers to maintain humidity.
     - Seal the plates or use parafilm to minimize evaporation.

9. **Problem: Cross-contamination between strains**
   - **Possible causes**
     - Handling multiple strains without proper segregation.
     - Use of non-sterile tools for inoculation.
   - **Solutions**
     - Work with one strain at a time and sterilize tools (e.g. inoculation loops) between strains.
     - Label plates and tubes clearly to avoid confusion.
10. **Problem: Difficulties in measuring growth**
    - **Possible causes**
      - Visual observation of fungal growth can be subjective.
      - Dense or pigmented cultures can make measurement difficult.
    - **Solutions**
      - Use a spectrophotometer to measure the optical density (OD) of liquid cultures.
      - Use metabolic assays (*e.g.*, XTT (2,3-Bis-(2-Methoxy-4-Nitro-5-Sulfophenyl)) or resazurin) for more accurate quantification.
      - For solid cultures, use colony-forming unit (CFU) counts or dry weight measurements.
11. **Problem: Hazardous exposure to pathogens**
    - **Possible causes**
      - Inadequate personal protective equipment.
      - Accidental spillage or exposure.
    - **Solutions**
      - Always wear appropriate PPE (lab coat, gloves, safety goggles, and mask/respirator if required).
      - Disinfect spills immediately and report them according to laboratory biosafety protocols.
      - Work with pathogenic fungi in a BSL-2 or higher laboratory under proper containment.
12. **Problem: Experimental variability**
    - **Possible causes**
      - Inconsistent preparation of the inoculum.
      - Variability of environmental conditions.
    - **Solutions**
      - Standardize inoculum preparation and growth conditions.
      - Perform experiments in replicates to ensure reproducibility.
      - Use controls in each experiment to account for variability.

2.2 Cultivation of Fungal Pathogens

Cultivation of fungal pathogens is a fundamental technique in mycology, pathology and microbiology that enables the study of fungal biology, pathogenicity, host-pathogen interactions and disease management. Successful cultivation of fungal pathogens requires an understanding of their specific growth requirements, selection of appropriate media and optimization of environmental conditions. The following is a detailed guide to the cultivation of fungal pathogens.

2.2.1 Purpose of Cultivation of Fungal Pathogens

The cultivation of fungal pathogens serves several purposes, including:

- Isolation of pathogens from infected host tissue or environmental samples.
- The propagation of fungi for morphological, physiological and genetic studies.
- Performing pathogenicity tests and antifungal susceptibility tests.
- Production of spores or mycelium for research and industrial applications.

2.2.2 Isolation of Fungal Pathogens

- The first step in cultivation is the isolation of fungal pathogens from infected material, such as plant tissue, soil or clinical samples. The process usually involves the following steps:
  - **Surface sterilization:** the host tissue is treated with disinfectants (*e.g.* sodium hypochlorite or ethanol) to remove surface contaminants.
  - **Plating on selective media:** Sterile tissue fragments are placed on a suitable growth medium to encourage fungal growth while suppressing bacteria.

2.2.3 Selection of Growth Media

Fungal pathogens have different nutritional and environmental requirements, necessitating suitable media. Commonly used media include:

- **PDA:** A universal medium that supports the growth of most fungi.
- **YEPD:** Optimized for yeasts, particularly pathogenic species,
- **Selective media:** The addition of antibiotics such as chloramphenicol or streptomycin helps to suppress bacterial growth during fungal isolation.

2.2.4 Environmental Conditions for Cultivation

Environmental factors play a crucial role in the growth and development of fungal pathogens:

- **Temperature:** most fungi thrive at 20–30 °C. Pathogenic fungi, especially those that infect humans, may require higher temperatures (35–37 °C).

- **Humidity:** High humidity is generally required for fungal growth. Humidity chambers can be used to maintain optimum humidity.
- **pH:** Acidic conditions (pH 4–6) favor fungal growth, although some species prefer a neutral pH.
- **Light:** Some fungi show photomorphogenic reactions and require a light-dark cycle for sporulation.

2.2.5 Cultivation Techniques

Cultivation methods for fungi are generally divided into two main categories: Plate culture and liquid culture. These methods are summarized in the following subsections:

- **Plate cultures:** plate cultures generally use agar as a solidifying agent. Some types, such as YEPD agar and PDA, are nutrient-rich. At the same time, minimal agars require the addition of essential nutrients such as carbon and nitrogen sources. YEPD is generally preferred for the cultivation of yeasts, while PDA is commonly used for filamentous fungi. The following protocol describes the steps for growing and maintaining fungal cultures on solid media:
  - Suspend the required amount of YEPD agar/PDA in 1 L of distilled water in a sterile glass flask and seal it with a cotton stopper.
  - Heat to boiling while stirring to dissolve all ingredients completely.
  - Autoclave for 15 min at 121 °C and 15 psi.
  - Pour about 25–30 mL of the agar medium onto sterile Petri dishes and allow it to settle
  - Spread the cells onto the agar plates and incubate at 30–35 °C for 24–48 h if yeast. Filamentous fungi are usually grown by cutting off the old culture block (5 mm) and placing it on the new plate or agar plate and incubating at 25–30 °C for 5–7 days, depending on the fungal species.
- **Liquid cultures**: Liquid cultures are generally carried out in conical flasks on a shaker. In the laboratory, fungal shake cultures are usually grown in 50 mL medium in 250 mL Erlenmeyer flasks, but cultivation can also be scaled up or down. The following protocol outlines the steps for culturing and maintaining fungal cultures in liquid media:
  - Suspend the required YEPD/PDA broth in 1 L of distilled water in a sterile glass flask sealed with a cotton plug.
  - Autoclave for 15 min at 121 °C and 15 psi.
  - Inoculate the fungal cultures into detergent-free tubes/bottles containing the prepared liquid medium. The medium should not exceed one-third of the volume of the culture tube or one-fifth of the total volume of the flask.
  - Vortex the contents briefly to disperse the cells.
  - Grow the culture in a shaking incubator at 150 rpm at 30–35 °C for 24–48 h in the case of yeast and 25–30 °C for 5–7 days, depending on the fungal species.

2.2.6 Special Cultivation Requirements

Some fungal pathogens, such as obligate biotrophs, require special hosts or dual cultures with compatible organisms for successful growth. These fungi are cultivated in plant tissues or in special co-cultures.

2.2.7 Dealing with Contamination

Contamination is a common challenge with fungal cultures. Aseptic techniques, the use of sterilized equipment and selective media help to reduce contamination. Contaminated cultures should be discarded after proper sterilization to prevent environmental spread.

2.2.8 Quality Control and Monitoring

Regular monitoring of fungal cultures for morphological integrity and purity is essential. Contamination, changes in colony morphology or unusual growth patterns may indicate problems in the cultivation process.

2.2.9 Challenges and Solutions

- **Slow-growing fungi:** Longer incubation times may be required for slow-growing species.
- **Obligate pathogens:** Developing artificial cultivation techniques or surrogate host systems is necessary.
- **Pathogenic variability**: Different strains of the same species may require customized media and conditions.

2.2.10 Applications of Fungal Cultures

Cultured fungal pathogens are essential for:

- Disease diagnosis and identification.
- Investigation of fungal life cycles, spore production, and infection mechanisms.
- Testing fungicides and biocontrol agents.
- Molecular and biochemical studies to understand pathogen virulence.

Therefore, cultivating fungal pathogens is a meticulous process that requires careful selection of media, optimization of environmental conditions and adherence to aseptic protocols. By mastering these techniques, researchers can effectively cultivate fungi, paving the way for advances in fungal biology, disease management and industrial applications.

2.2.11 Troubleshooting

1. **Problem: No fungal growth observed**
   - **Possible causes**
     - Non-viable fungal inoculum.
     - Incorrect or suboptimal growth medium.
     - Unsuitable incubation conditions.
   - **Solutions**
     - Check the viability of the fungal strain by cultivating a fresh inoculum from a strain.
     - Use the appropriate medium for the fungal species.
     - Optimize the incubation conditions specific to the pathogen (temperature, humidity and pH). For example,
       - 25–30 °C for environmental fungi.
       - 35 °C for human pathogenic fungi such as *Candida* sp.
2. **Problem: Contamination in cultures**
   - **Possible causes**
     - Non-sterile techniques or materials.
     - Contaminated medium or fungal strain.
     - Prolonged contact of cultures with air.
   - **Solutions**
     - Autoclave all media, tools, and containers before use.
     - Work in a fume hood with laminar airflow or a biological safety cabinet to minimize contamination.
     - Use antifungal or antibacterial agents in the medium to suppress unwanted microbial growth.
     - Avoid leaving culture plates or bottles open for extended periods of time.
3. **Problem: Poor or uneven growth**
   - **Possible causes**
     - Insufficient nutrients in the medium.
     - Uneven distribution of the fungal inoculum
     - Overfilling on solid media.
   - **Solutions**
     - Use a nutrient-rich medium that is specific for the fungal pathogen.
     - Ensure even mixing of the fungal inoculum in liquid media or spread the inoculum evenly on solid media.
     - Inoculate fewer spores or cells per plate to avoid overcrowding.
4. **Problem: Overgrowth of fungal cultures**
   - **Possible causes**
     - High inoculum density.
     - Excessively long incubation time.

- **Solutions**
  - Adjust the concentration of fungal inoculum to the recommended levels.
  - Monitor cultures regularly and stop incubation when optimum growth is reached.
  - Use appropriate dilution techniques to separate colonies on solid media.

5. **Problem: Formation of clumps in liquid cultures**
   - **Possible causes**
     - Aggregation of fungal spores or cells.
     - Insufficient shaking during incubation.
   - **Solutions**
     - Add a dispersing agent such as Tween-20 (0.01–0.1%) to liquid cultures to prevent clumping.
     - Use an orbital shaker for continuous shaking (120–200 rpm, depending on species).
     - Filter the inoculum through sterile gauze or mesh to break up clumps before inoculation.
6. **Problem: Inconsistent spore formation**
   - **Possible causes**
     - Insufficient exposure to light or temperature.
     - Nutrient deficiencies in the medium.
   - **Solutions**
     - Use sporulation-promoting media such as PDA.
     - Provide alternating light-dark cycles if the species requires light for sporulation.
     - Incubate at temperatures favorable for sporulation (usually 20–25 °C for many fungi).
7. **Problem: Fungal cultures dry out**
   - **Possible causes**
     - Evaporation due to low humidity.
     - Loose or improperly sealed plates or containers.
   - **Solutions**
     - Incubate the cultures in a humidified environment or seal the plates with parafilm.
     - Use moist chambers or place moist paper towels in the incubator to maintain humidity.
     - Regularly check the plates and subculture to fresh plate if media/agar has reduced.
8. **Problem: Difficulties in isolating pure cultures**
   - **Possible causes**
     - Mixed cultures due to contamination.
     - Poor streaking techniques.
   - **Solutions**
     - Perform streak plating on selective media to isolate single colonies.
     - If necessary, use antibiotics in the medium to suppress bacterial growth.

  - Pick individual colonies with sterile tools and subculture them on fresh plates.

9. **Problem: Slow growth or prolonged lag phase**
   - **Possible causes**
     - Dormant or stressed fungal spores/cells.
     - Insufficient aeration in liquid cultures.
   - **Solutions**
     - Revive the strain by subculturing it on fresh plates or in fresh medium before use.
     - Aerate the liquid cultures appropriately with orbital shakers or sparring systems.
     - Check the pH of the medium and adjust it to the fungal species (typically pH 5–6 for most fungi).
10. **Problem: Changed morphology of the colonies**
    - **Possible causes**
      - Mutations due to prolonged storage or repeated subculturing.
      - Environmental stress (*e.g.*, pH value, temperature fluctuations).
    - **Solutions**
      - Use freshly revived strains from a stock culture.
      - Minimize the number of subcultures to reduce the risk of genetic drift.
      - Standardize environmental conditions for all experiments

2.3 Preservation of Fungal Cells

The preservation of fungal cells is a fundamental technique in microbiology and mycology. It enables the long-term storage of fungal strains while maintaining their viability, genetic stability and phenotypic properties. Appropriate preservation methods are important to maintain reference strains, support research and facilitate industrial applications. The techniques, challenges and considerations for preserving fungal cells are explained in detail below.

2.3.1 Importance of Fungal Cell Preservation

The preservation of fungal cells is crucial for several reasons, including:

- **Continuity in research:** preserved cells allow researchers to conduct experiments over longer periods of time without having to re-isolate strains.
- **Genetic stability:** minimizing genetic drift during storage ensures reliable and reproducible experimental results.
- **Industrial applications:** Preserved fungi are used to produce enzymes, pharmaceuticals and biofuels.
- **Preservation:** he preservation of fungal strains helps to maintain genetic diversity and protect rare or endangered species.

2.3.2 Methods for the Preservation of Fungal Cells

- **Short-term preservation**
  - **Agar slants:** Fungal cultures are grown on agar slants, sealed to prevent desiccation and stored at 4–10 °C. This method is suitable for storing cultures for weeks to months.
  - **Storage in sterile water:** Fungal spores or mycelial fragments are suspended in sterile distilled water and stored at room temperature or in a refrigerator. This simple and inexpensive method can maintain viability for up to 1 year.
- **Long-term preservation**
  - **Cryopreservation:** Fungal cells are suspended in cryoprotectants (*e.g.*, glycerol, dimethyl sulfoxide (DMSO)) to prevent the formation of ice crystals and stored in liquid nitrogen (−196 °C) or ultra-low temperature freezers (−80 °C).
    - **Advantages:** Preserves genetic and physiological stability for decades.
    - **Applications:** Widely used to preserve pathogenic and industrially important fungi.
  - **Lyophilization (Freeze-Drying):** Fungal cells are suspended in a protective medium (*e.g.*, skim milk or sucrose solution), frozen, and subjected to vacuum drying. The dried cells are stored at room temperature in sealed vials.
    - **Advantages:** Long shelf life and easy transportation.
    - **Applications:** Commonly used to preserve fungal spores and certain yeasts.
  - **Silica gel preservation:** Fungal spores are mixed with sterile silica gel and stored in airtight containers.
    - **Advantages:** Inexpensive and suitable for long-term storage at room temperature.
    - **Applications:** Best suited for spore-forming fungi.
- **Specialized techniques**
  - **Mineral oil overlay:** Fungal cultures are grown on agar plates or slants, covered with sterile mineral oil, and stored at 4 °C. The oil prevents dehydration and contamination.
  - **Encapsulation in calcium alginate beads:** Fungal cells are immobilized in calcium alginate beads, dried, and stored. This method is less common but effective for certain applications.

Preservation in Glycerol Vials

- **Material needed**
  - Fungal culture (pure and actively growing)
  - Sterile cryovials or storage tubes (2 mL recommended)
  - Sterile glycerol stock solution
  - Sterile water or suitable fungal growth medium
  - Pipettes and sterile pipette tips
  - Vortex mixer
  - −80 °C freezer for storage

- **Procedure**
  - Grow the fungal culture in an appropriate growth medium (solid or liquid) until it reaches a robust growth phase.
    - For liquid cultures: Ensure the culture is at its exponential or stationary growth phase.
    - For solid cultures: Scrape a sufficient amount of mycelium or spores.
  - Prepare sterile glycerol by diluting stock glycerol to the desired final concentration (typically 30% v/v). Autoclave the diluted glycerol solution or filter-sterilize it using a 0.22 μm filter.
  - Harvest the fungal culture.
    - For liquid cultures: Centrifuge the culture at 3000–5000×*g* for 5–10 min to pellet the mycelium or spores. Discard the supernatant.
    - For solid cultures: Gently scrape the fungal mycelium or spores using a sterile loop or scalpel and transfer them to a sterile container.
  - Suspend about 500 μL of fungal biomass (mycelium or spores) in 500 μL of sterile 30% glycerol. Mix thoroughly by vortexing gently.
  - Transfer 1 mL of the glycerol-fungal suspension into sterile cryovials.
  - Label the cryovials with essential information, such as fungal species/strain, date, and storage concentration.
  - Place the cryovials in a −80 °C freezer for long-term storage.
- **Notes**
  - Glycerol concentration: Adjust based on the species; some fungi tolerate higher concentrations better.
  - Aseptic technique: Maintain sterility throughout the process to prevent contamination.
  - Viability testing: Periodically check the viability of stored cultures by reviving and culturing them.
  - Alternatives for preservation: Lyophilization and mineral oil overlay are alternative methods for preserving fungal cultures.

Preservation by Freeze Drying or Lyophilization

- **Material needed**
  - Fungal culture (pure and actively growing)
  - 10% sterile skimmed milk
  - Sterile lyophiles
  - Sterile pipettes and tips
  - −40 °C and 4 °C freezer for storage
- **Procedure**
  - Grow fungal culture on an appropriate medium to reach a high biomass density. Harvest the spores or mycelium.
  - Suspend spores or mycelial fragments in a cryoprotectant solution (*e.g.*, sterile skim milk) to prevent freezing damage.

  - Dispense 250–300 μL of suspension in labeled lyophiles and rapid freezing is achieved by putting the sample at −40 °C for 30 min.
  - Then immediately put in the lyophilizer (vacuum pump and condenser which maintains −20 to −30 °C).
  - Seal the lyophils with a sealing torch after the cultures have been lyophilized (approx. 6–7 h).
  - Allow the frozen suspension to dry completely by keeping it for 5–6 h under vacuum.
  - Seal the lyophilizes properly with a sealing torch and store at 4 °C. The lyophilized cultures can be preserved for more than 20 years [1].
- **Notes**
  - Cryoprotectant selection: Protecting the cellular structure during freezing is critical.
  - Species variability: Some fungi (*e.g.*, yeasts or fastidious fungi) may require optimization of protocols.
  - Rehydration protocol: Improper rehydration can reduce viability.

2.3.3 Challenges in the Preservation of Fungal Cells

- **Cell damage:** Freezing, desiccation, or suboptimal storage conditions can damage fungal cells.
- **Metabolic activity:** An active metabolism during storage can consume nutrients or cause genetic changes.
- **Pathogenicity:** Handling pathogenic fungi requires adherence to biosafety protocols to prevent contamination and the spread of disease.

2.3.4 Factors Influencing the Success of Preservation

- **Species characteristics:** Spore-forming fungi are generally easier to preserve than obligate biotrophs.
- **Storage conditions:** Temperature, humidity, and oxygen content significantly affect the viability of fungi.
- **Cryoprotectants and additives:** The choice of preservatives influences the survival of cells survival during freezing and drying

2.3.5 Applications of Preserved Fungal Cells

- **Pathogen research:** Preserved pathogens are crucial for disease diagnostics and testing control measures.
- **Industrial production:** Fungal strains are used in fermentation processes to produce enzymes, antibiotics, and organic acids.

- **Biodiversity studies:** Preserved fungi contribute to the understanding of the ecological role and conservation of fungal germplasm.

Therefore, preserving fungal cells is a cornerstone of fungal biology and biotechnology and enables the sustainable use of fungal resources in research, industry and biodiversity conservation. By using appropriate techniques and storage conditions, researchers can ensure the long-term viability and stability of fungal strains, enabling advances in various scientific and industrial fields. Continued innovation in preservation methods will further enhance the ability to effectively store and utilize fungal resources.

2.3.6 Troubleshooting

1. **Problem: Loss of viability after thawing**
   - **Possible causes**
     - Unsuitable concentration of cryoprotectant.
     - Unsuitable freezing or thawing protocols.
     - Repeated freeze-thaw cycles.
   - **Solutions**
     - Use suitable cryoprotectants (*e.g.*, 15–20% glycerol, 10% DMSO) to prevent the formation of ice crystals
     - Freeze the cells slowly (*e.g.* at −1 °C/min) using a controlled freezing device or isopropanol container before transferring the cells to −80 °C or liquid nitrogen.
     - Thaw the cells quickly in a water bath at 37 °C and use them immediately so that they are not exposed to room temperature for too long.
     - Aliquot fungal stocks into smaller volumes to avoid repeated freeze-thaw cycles.
2. **Problem: Low recovery rate**
   - **Possible causes**
     - Insufficient cell density before freezing.
     - Damage during storage or revival.
     - Contamination in frozen stocks.
   - **Solutions**
     - Ensure that the fungal cultures are in the exponential phase before harvesting for freezing.
     - Set the cell/spore concentration to an optimal level (*e.g.* 10^6–10^7 cells/mL).
     - Use sterile techniques when preparing and storing stocks to avoid contamination.
3. **Problem: Contamination in preserved stocks**
   - **Possible causes**
     - Non-sterile handling during preparation.
     - Contaminated cryoprotectants or containers.
     - Improper sealing of storage vials.

- **Solutions**
  - Use sterilized cryovials and cryoprotectants for preservation.
  - Work in a laminar flow hood or biological safety cabinet to maintain sterility.
  - Ensure that the cryovials are tightly sealed to prevent contamination during storage.
  - Inspect frozen stocks regularly for contamination by thawing and culturing the samples.

4. **Problem: Loss of genetic stability**
   - **Possible causes**
     - Frequent subcultivation before freezing.
     - Long-term storage without adequate cryoprotection.
   - **Solutions**
     - Minimize subculturing to reduce genetic drift; preserve strains as soon as possible after isolation.
     - Use long-term storage methods such as freezing at −80 °C or in liquid nitrogen (−196 °C).
     - Revive the strains regularly and confirm their identity using molecular or phenotypic methods.
5. **Problem: Lack of formation of viable stocks**
   - **Possible causes**
     - Poor quality or incorrect preparation of the cryoprotectant.
     - Inefficient mixing of the cells with the cryoprotectant.
   - **Solutions**
     - Prepare fresh cryoprotectants and check their compatibility with the fungal strain.
     - Mix the fungal cultures with the cryoprotectants thoroughly but carefully to avoid mechanical damage.
     - Test different cryoprotectants or concentrations to determine the optimal conditions for each fungus.
6. **Problem: Dehydration of freeze-dried cultures**
   - **Possible causes**
     - Improper sealing of the storage vials.
     - Storage in a humid environment.
   - **Solutions**
     - Ensure that lyophilized fungal stocks are sealed in airtight containers, preferably under vacuum or with desiccants.
     - Store freeze-dried cultures in a low-humidity environment at 4 °C or lower.
     - Avoid opening the vials unnecessarily to prevent moisture ingress.
7. **Problem: Difficulties in reviving lyophilized cultures**
   - **Possible causes**
     - Damage to the cells during freeze-drying.
     - Incorrect rehydration procedure.

- **Solutions**
  - Use cryoprotectants such as glycerol or DMSO before lyophilization to protect fungal cells.
  - Carefully rehydrate lyophilized cells with sterile water or culture medium at room temperature.
  - Place rehydrated cultures on a nutrient-rich medium to facilitate recovery.

8. **Problem: Ice crystal damage during freezing**
   - **Possible causes**
     - Rapid freezing without controlled cooling.
     - Incorrect use of a cryoprotectant.
   - **Solutions**
     - Use a controlled rate freezing method to prevent intracellular ice formation.
     - Add cryoprotectants such as glycerol or DMSO to protect cells from freezing damage.
     - Ensure that the cryoprotectant is evenly mixed with the culture before freezing.
9. **Problem: Difficulties in preserving spore-forming fungi**
   - **Possible causes**
     - Improper harvesting or handling of spores.
     - Unsuitable storage conditions.
   - **Solutions**
     - Harvest the spores at their peak viability using sterile methods.
     - Store the spores dry or in a cryoprotectant, depending on the type of fungal species.
     - Freeze the spores in liquid nitrogen for long-term preservation or store them lyophilized.
10. **Problem: Difficulties in the preservation of non-spore-forming fungi**
    - **Possible causes**
      - Greater sensitivity to freezing and thawing compared to spore-forming fungi.
    - **Solutions**
      - Use higher concentrations of cryoprotectants for non-spore-forming fungi.
      - Test alternative preservation methods such as encapsulating cells in agar beads before freezing.
      - Freeze fungal cultures directly in liquid medium containing cryoprotectant to improve viability.
11. **Problem: Accidental thawing during storage**
    - **Possible causes**
      - Malfunction of the freezer or power failure.
      - Poor monitoring of storage conditions.
    - **Solutions**
      - Use freezers with alarm systems to alert users in the event of a malfunction.
      - Monitor storage temperature with backup temperature loggers.

- Immediately move stock to another storage facility (e.g. another freezer) when it defrosts.

12. **Problem: Problems with long-term storage stability**
    - **Possible causes**
      - Decomposition of the cryoprotector over time.
      - Insufficient storage temperature.
    - **Solutions**
      - Replace the cryoprotectants if the stocks are stored for a long period of time
      - Store at ultra-low temperatures (−80 °C or in liquid nitrogen) to ensure long-term stability.
      - Carry out regular viability checks and prepare new stocks if necessary.

2.4 Resuscitation of Fungal Cells

The revival of fungal cultures is a crucial step in microbiology and mycology, ensuring that preserved fungal strains are returned to an active and viable state for research, industrial applications or diagnostics. To preserve the physiological, genetic and morphological integrity of fungal cells, appropriate resuscitation techniques are essential. The following is a comprehensive guide to the process, considerations and challenges of resuscitating fungal cultures [3].

2.4.1 Importance of Fungal Cultures Revival

- Use the preserved strains for experiments or industrial processes.
- Ensure the continuation of long-term studies without re-isolation of fungi.
- Confirm the viability and purity of preserved cultures.
- Restore rare or valuable fungal strains to preserve biodiversity conservation efforts.

2.4.2 General Steps in the Revival Process

- **Preparation**
  - **Assess preservation method:** Determine the preservation technique used (*e.g.* lyophilization, cryopreservation, or silica gel storage) as it determines the revival procedure.
  - **Check the storage integrity:** inspect the storage vials or containers for signs of contamination, physical damage or compromised seals.
- **Rehydration**
  - Add sterile distilled water or a nutrient-rich medium to the preserved fungal sample.
  - Allow the sample to sit for 15–30 min to ensure gradual rehydration of the cells.

- Mix carefully to avoid damaging the cells or spores.
- **Inoculation onto growth media**
  - Transfer the rehydrated fungal material to a suitable solid or liquid growth medium (PDA/YEPD) using sterile techniques.
  - For spore-forming fungi, you can sprinkle or spread a small amount of the spore suspension on the surface of the medium.
  - For non-spore-forming fungi, transfer mycelial fragments directly onto the medium.
- **Incubation:** Incubate the inoculated medium under optimal conditions for the fungal species:
  - **Temperature:** Most fungi grow well at 20–30 °C, but human pathogenic fungi may require 35–37 °C.
  - **Humidity:** Provide high humidity to promote fungal growth, especially for obligate pathogens.
  - **Light conditions:** Provide light-dark cycles if the fungal species requires photomorphogenic cues for sporulation or development.
- **Monitoring growth**
  - Check the culture regularly for signs of fungal growth, such as the emergence of mycelium or spores.
  - Observe the morphology of the colony to ensure that it matches the expected characteristics of the preserved strain.

2.4.3 Quality Control

After revival, it is important to confirm the identity and integrity of the culture:

- **Morphological examination:** compare the colony morphology, spore structure, and hyphal characteristics with the reference descriptions.
- **Molecular identification:** Use DNA-based techniques such as PCR or sequencing of the Internal Transcribed Spacer (ITS) region to verify the fungal species.
- **Pathogenicity testing:** For pathogenic fungi, evaluate their virulence and host-specific characteristics to confirm that they remain intact.

2.4.4 Challenges in the Revival of the Fungus

- **Loss of viability:** Fungi that have been preserved under suboptimal conditions cannot be revived. Adaptation of rehydration or incubation methods can sometimes help to recover such cultures.
- **Contamination:** Preserved cultures may become contaminated during storage or revival. Careful aseptic techniques and repeated transfers to fresh media can mitigate this problem.

- **Physiological changes:** Long-term preservation may result in phenotypic changes, such as reduced sporulation or altered colony morphology. Maintaining optimal revival conditions minimizes these effects.
- **Obligate pathogens:** Some fungi require living hosts or specialized media for revival, posing additional challenges. Techniques like dual cultures with compatible organisms or inoculation into host tissues can be employed.

2.4.5 Specialized Revival Techniques

- **Cryopreserved cultures**
  - Thaw the cryovial quickly by placing it in a water bath at 37 °C for 1–2 min.
  - Transfer the contents to a recovery medium immediately to avoid prolonged exposure to room temperature.
- **Lyophilized cultures**
  - Rehydrate the freeze-dried pellet by adding sterile water or nutrient broth.
  - Mix gently to ensure even hydration before inoculating on solid media.
- **Silica gel-stored cultures**
  - Extract spores from silica gel granules using sterile water.
  - Inoculate the spore suspension onto the appropriate medium.

2.4.6 Applications of Revived Fungal Cultures

- **Pathogen studies:** Revived cultures are used to study disease mechanisms and develop antifungal strategies.
- **Industrial processes:** Many industries depend on fungi for producing enzymes, organic acids, antibiotics, and biofuels.
- **Biodiversity conservation:** Revived fungi contribute to understanding ecological roles and maintaining germplasm collections.
- **Educational purposes:** Preserved and revived cultures are essential for teaching and training in mycology.

The revival of fungal cultures is a delicate but essential process that ensures the usability and integrity of the preserved strains. By following precise protocols and maintaining optimal conditions during rehydration, inoculation and incubation, researchers can successfully recover fungal cultures. This capability supports various applications in research, industry and conservation, making the revival process a cornerstone of fungal biology and applied mycology.

2.4.7 Troubleshooting

1. **Problem: Low viability or no growth after revival**

- **Possible causes**
  - Prolonged storage causes cell death.
  - Ineffective or inappropriate cryoprotectant.
  - Damage during freezing or thawing processes.
- **Solutions**
  - Use fresh stocks if viability is too low.
  - Ensure cryoprotectants like 15–20% glycerol or 10% DMSO are used correctly.
  - Thaw frozen stocks rapidly (*e.g.*, in a 37 °C water bath) to minimize ice crystal damage.
  - Plate or culture cells on nutrient-rich medium like PDA to promote growth.

2. **Problem: Contamination in revived cultures**
   - **Possible causes**
     - Non-sterile handling during revival.
     - Contaminated stock cultures.
     - Poor sealing or storage conditions.
   - **Solutions**
     - Work in a sterile environment, such as a laminar airflow hood or biological safety cabinet.
     - Use antibiotics or antifungals in the medium if contamination is suspected (e.g., chloramphenicol for bacterial contamination).
     - Check frozen stocks periodically for contamination and discard compromised vials.
3. **Problem: Clumping of cells or spores**
   - **Possible causes**
     - Aggregation during freezing or lyophilization.
     - Inadequate mixing after thawing.
   - **Solutions**
     - Disperse clumps by gentle vortexing or pipetting up and down.
     - Add dispersing agents like 0.01–0.1% Tween-20 to the suspension.
     - Pass the suspension through sterile gauze or a mesh filter to remove clumps.
4. **Problem: Delayed growth or extended lag phase**
   - **Possible causes**
     - Stressed or damaged cells from storage.
     - Insufficient nutrients in the recovery medium.
     - Unsuitable incubation conditions (*e.g.*, temperature, humidity).
   - **Solutions**
     - Use a recovery medium enriched with nutrients (e.g., YEPD).
     - Incubate under optimal conditions for the fungal species (*e.g.*, 25–30 °C for environmental fungi, 37 °C for yeast pathogens).
     - Allow extra time for growth; some fungi may require 48–72 h to recover fully.
5. **Problem: Inconsistent colony morphology**
   - **Possible causes**
     - Mutations or genetic drift during long-term storage.

   – Stress responses during freezing or lyophilization.
- **Solutions**
   – Confirm fungal identity using molecular or phenotypic methods.
   – Minimize the number of subcultures before storage to reduce the risk of genetic drift.
   – Revive from multiple stocks to check for consistency.

6. **Problem: Loss of sporulation ability**
   - **Possible causes**
     – Damage to spore-forming structures during storage.
     – Suboptimal revival conditions.
   - **Solutions**
     – Incubate on sporulation-inducing media.
     – Ensure light or temperature conditions conducive to sporulation (*e.g.*, light-dark cycles if required).
     – Use freshly revived cultures for sporulation studies.
7. **Problem: Incomplete rehydration of lyophilized cells**
   - **Possible causes**
     – Incorrect rehydration process.
     – Rapid or uneven rehydration causing cellular stress.
   - **Solutions**
     – Rehydrate lyophilized cells gently with sterile water or an appropriate liquid medium at room temperature.
     – Allow the suspension to sit undisturbed for a few minutes to ensure complete rehydration.
     – Plate the rehydrated cells onto nutrient-rich media immediately.
8. **Problem: Difficulty recovering specific fungal strains**
   - **Possible causes**
     – Strain-specific growth requirements not being met.
     – Loss of viability due to unsuitable storage conditions.
   - **Solutions**
     – Optimize medium and conditions specific to the fungal strain.
     – Use selective media if the strain has distinct growth markers.
     – Check storage protocols to prevent future viability issues.
9. **Problem: Poor growth in liquid media**
   - **Possible causes**
     – Incorrect inoculum concentration.
     – Insufficient aeration or agitation.
   - **Solutions**
     – Adjust the initial inoculum concentration to the appropriate range (*e.g.*, 10^3–10^6 CFU/mL).
     – Use an orbital shaker (120–200 rpm) to provide adequate aeration.
     – Supplement the medium with additional nutrients if necessary.
10. **Problem: Revival failure in cryopreserved fungal cells**
   - **Possible causes**
     – Improper freezing or thawing protocol.

  – Inadequate mixing of cryoprotectant with the culture.
- **Solutions**
  – Ensure slow freezing (*e.g.*, −1 °C/min) before transferring to −80 °C or liquid nitrogen.
  – Mix cryoprotectant thoroughly with fungal cultures before freezing.
  – Avoid extended storage in frost-free freezers that experience temperature fluctuations.

11. **Problem: Formation of non-viable biomass**
    - **Possible causes**
      – Overgrowth of non-reproductive hyphae.
      – Stress responses from storage conditions.
    - **Solutions**
      – Subculture onto solid media to isolate viable colonies.
      – Select actively growing regions of the colony for subsequent inoculations.
12. **Problem: Inconsistent results across revivals**
    - **Possible causes**
      – Variability in storage conditions.
      – Differences in handling techniques.
    - **Solutions**
      – Standardize revival protocols across experiments.
      – Use replicate stocks to ensure consistency.
      – Monitor storage temperatures and conditions regularly.

References

1. Gupta S, Saxena A (2019) Experimental comparison: methods for preserving fungal cultures. Curr Res Environ Appl Mycol 9:208–212. https://doi.org/10.5943/cream/9/1/18
2. Moore GE, Gerner RE, Franklin HA (1967) Culture of normal human leukocytes. JAMA 199(8):519–524. https://doi.org/10.1001/jama.1967.03120080053007
3. Yang SZ, Peng LT (2023) Significance of the plasma membrane H^+-ATPase and V-ATPase for growth and pathogenicity in pathogenic fungi. In: Gadd GM, Sariaslani S (eds) Advances in applied microbiology. Academic Press, pp 31–53. https://doi.org/10.1016/bs.aambs.2023.07.001

Counting Colony Forming Units (CFU) to Estimate Viable Fungal Cells in a Sample

3

Abstract

The colony forming units (CFU) counting assay is a widely used technique for estimating the number of viable fungal cells in any sample. This method takes advantage of the unique ability of individual fungal cells to proliferate and form visible colonies on agar plates, providing a reliable measure of their reproductive capacity (Meyer et al, Nat Microbiol 8:2304–2314, 2023). The procedure begins with cultivating a fungal culture in a suitable liquid medium. Serial dilutions of the sample are then performed to ensure that the number of colonies formed on the agar plates remains within the optimal range of 30–300 per plate, which is essential for accurate counting. Once the dilutions are prepared, the diluted samples are plated onto specially formulated agar media and incubated at a temperature conducive to fungal growth. During the incubation period, the fungal cells multiply and form distinct colonies. After a predetermined incubation period, the colonies are counted and the concentration of viable fungal cells is expressed as CFU/mL, which is calculated based on the dilution factor (DF) and the volume of the plated sample. The CFU assay is critical in many applications, including fungal susceptibility testing, where it helps to determine the efficacy of antifungal drugs against specific fungal strains (Sahu SR et al, Bio Protoc 13:e4872, 2023). It is also central to pathogenicity assessment and helps researchers understand the virulence of fungal pathogens. In addition, the CFU assay is crucial for quantifying the microbial load of environmental samples, contributing to studies on biodiversity and ecosystem health. Its simplicity, accuracy and adaptability to different fungal species make it an indispensable tool in fungal microbiology, enabling advances in research and clinical diagnostics. Overall, the CFU counting assay remains a cornerstone of mycology, enabling researchers to elucidate complex interactions within fungal communities and their environments (Sieuwerts et al, Lett Appl Microbiol 47:275–278, 2008).

R. Pasrija, D. Kumari, *Protocols for Studying Pathogenic Fungi*,
https://doi.org/10.1007/978-981-96-7047-5_3

Keywords

Colony-forming unit · Fungal cells · Agar · Visible · Quantification

Introduction In this chapter, the protocols from culture preparation to the performance of the colony forming unit (CFU) count test are described in detail. The following topics are covered as follows:

3.1. Materials Required
3.2. Preparations of Media and Buffers
3.3. Procedure
3.4. Tips and Troubleshooting

3.1 Materials Required

3.1.1 Fungal Sample and Medium

- **Fungal sample**: Suspension with the fungal cells or spores to be quantified.
- **Growth medium**: Agar plates suitable for the fungal species, *e.g.* yeast extract peptone dextrose (YEPD) agar.

3.1.2 Materials for Dilution

- **Sterile saline solution or distilled water**: Sterile saline solution is used as a diluent or to prepare serial dilutions of the fungal suspension.
- **Dilution tubes**: Sterile test tubes or microcentrifuge tubes are used as dilution tubes.
- **Pipettes and tips**: Use sterile serological pipettes or micropipettes for accurate liquid handling.
- **Vortex mixer**: Use a vortex mixer to homogenize fungal suspensions.

3.1.3 Plating Materials

- **Sterile Petri dishes**: Use pre-cast agar plates for fungal growth.
- **Spreader**: Use a sterilized glass or metal rod to spread the fungal suspension evenly on the agar plate.

3.1.4 Sterilization and Safety Accessories

- **Bunsen burner or alcohol lamp**: For aseptic handling of samples and tools.
- **Gloves, lab coat, and eye protection**: For personal safety during the procedure.

3.1.5 Incubation Equipment

- **Incubator**: Set to the optimum temperature for the fungal species (usually 30 °C).

3.1.6 Counting Equipment

- **Colony counter (optional)**: For easier counting of the colonies.
- **Marker pen**: For labelling the plates with the sample data and DF.

3.1.7 Data Recording Tools

- **Notebook or spreadsheet**: For recording colony counts and calculating CFU/mL.
- **Calculator**: Calculate final CFU/mL values based on DF and plating volumes.

3.2 Preparations of Media and Buffers

3.2.1 Preparation of YEPD Media and Agar (See Chap. 2)

3.2.2 Preparation of Phosphate-Buffered Saline (PBS)

- **Materials required:** Sodium chloride (NaCl), potassium chloride (KCl), disodium hydrogen phosphate ($Na_2HPO_4{\cdot}2H_2O$), monopotassium phosphate (KH_2PO_4), distilled or deionized water, pH meter or pH indicator paper, 1 M hydrochloric acid (HCl) or sodium hydroxide (NaOH) for pH adjustment (if required), beaker or flask, stirring rod and magnetic stirrer (optional), *etc.*
- **Procedure:**
  - Accurately weigh the following chemicals: 8.0 g NaCl, 0.2 g KCl, 1.44 g $Na_2HPO_4{\cdot}2H_2O$ and 0.24 g KH_2PO_4 for 1 L PBS solution
  - Add the weighed chemicals to about 800 mL of distilled or deionized water in a beaker or flask. Then, stir the mixture with a magnetic stirrer (or by hand if necessary) until all components are completely dissolved.
  - Once the salts are dissolved, transfer the solution to a 1 L volumetric flask. Then add distilled or deionized water up to the 1 L mark. Ensure that the final volume is exactly 1 L.

- Use a pH meter to measure the pH of the solution which should be 7.4.
- If you need sterile PBS, sterilize the solution by autoclaving at 121 °C for 15–20 min or filter it through a 0.22 μm filter.
- Store the prepared PBS at room temperature or 4 °C, depending on what you want to use it for. PBS can be used for several weeks if stored properly, although it is usually recommended to use it within 1 month.

3.2.3 Preparation of the Saline Solution (0.85%)

- **Materials required:** NaCl, distilled or deionized water, beaker or flask, stirring rod and magnetic stirrer (optional), autoclave, *etc.*
- **Procedure**
  - To prepare a 0.85% saline solution, you need a concentration of 0.85 g NaCl per 100 mL of solution. For example, to prepare 1 L of 0.85% saline solution, you need: 0.85 g/100 mL × 1000 mL = 8.5 g NaCl.
  - Use a weighing balance to measure 8.5 g NaCl for 1 L of saline solution. Then, adjust the quantity proportionally if you prepare a different volume (*e.g.* you need 4.25 g NaCl for 500 mL).
  - Pour about 800 mL of distilled water into a beaker or flask. Then add the weighed NaCl to the water while stirring to dissolve it completely. You can use a magnetic stirrer or stir by hand.
  - Once the NaCl is completely dissolved, transfer the solution to a 1 L beaker and add distilled water to bring the total volume to 1 L.
  - If necessary, you can sterilize the saline solution by autoclaving it at 121 °C for 15–20 min.
  - Store the prepared 0.85% saline solution in a clean, tightly closed sterile container or bottle. Once sterilized, store at room temperature for short-term use or refrigerate it if you want to keep it for longer.

3.3 Procedure

3.3.1 Preparation of the Fungal Culture

- Streak all yeast strains of interest fresh from glycerol stock onto a sterile YEPD agar plate and incubate at 30 °C for 48 h.
- Inoculate a single colony of each strain into 5 mL of autoclaved YEPD broth and incubate overnight at 30 °C with shaking at 200 rpm. Centrifuge at 3000–4000 rpm for 5 min, discard the supernatant and resuspend the cell pellet in 1 mL of sterile PBS to wash the cells.

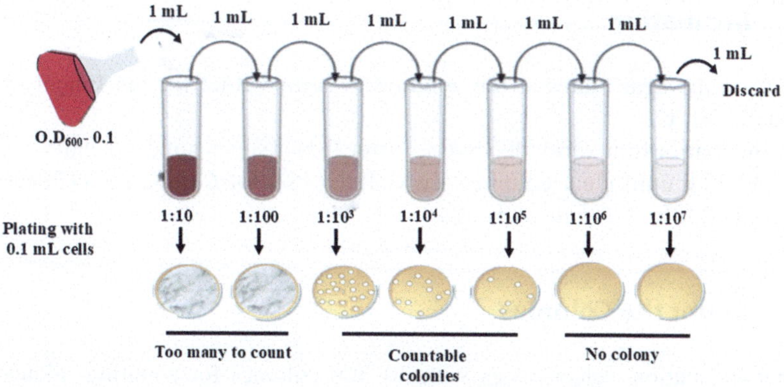

Fig. 3.1 Flow chart for counting colony forming units (CFU) in fungi

3.3.2 Serial Dilution

- Dilute the washed culture in the spectrophotometer in 0.85% saline to OD_{600} 0.1.
- Label sterile 15 mL or 50 mL tubes with the desired DF (*e.g.* 10^{-1}, 10^{-2}, 10^{-3}, *etc.*).
- Add 1 mL of the undiluted culture to the first tube containing 9 mL of saline (10^{-1} dilution). Shake briefly to mix.
- Transfer 1 mL from tube 10^{-1} to the next tube (10^{-2}) containing 9 mL of saline (as shown in Fig. 3.1). Shake briefly.
- Repeat this procedure for each additional dilution to obtain a series of dilutions (up to 10^{-6}, depending on the initial concentration).

3.3.3 Plating Out the Dilutions

For each dilution, add a certain volume (usually 100 μL) to agar plates:

- Pipette 100 μL of each dilution (starting with the most diluted sample) onto separate agar plates.
- Spread the sample evenly with a sterile spreader or sterile glass beads.
- If using glass beads, place them on the plate, shake gently in a circular motion to disperse the liquid, and then discard the beads in a sanitizing solution.
- Perform a duplicate or triplicate plate for each dilution to ensure accuracy.

3.3.4 Incubation

- Incubate the agar plates at the appropriate temperature for the fungal species (usually 30 °C).
- The incubation time varies depending on the fungal species and is usually between 48 and 72 h until clear colonies are visible (48 h for *Candida*/*Saccharomyces* sp. and 72 h for *Cryptococcus* sp).

3.3.5 Count the Colonies

- After incubation, select plates with 30–300 colonies for counting. Plates with less than 30 colonies are subject to greater statistical fluctuations, and plates with more than 300 colonies are too crowded.
- Count the number of colonies on each plate with a colony counter or by visual inspection.
- If the colonies are small or crowded, use a magnifying glass or stereomicroscope to make a more accurate count.

3.3.6 Calculation of CFU/mL

Calculate the CFU/mL of the original fungal suspension using the following formula:

$$\text{CFU / mL} = \text{Number of colonies / DF} \times \text{plated volume}\left(\text{in mL}\right)$$

Calculation example: If you get 2 colonies in 10^5 DF in 100 μL plated (Fig. 3.1):

$$\text{Colonies in } 100\,\mu L = 2\,\text{colonies, then } 1000\,\mu L \text{ cells would have}$$
$$= 2 / 100 \times 1000 \times \text{DF}\left(10^5\right) = 2 \times 10^6 \text{ CFU / mL.}$$

3.4 Tips and Troubleshooting

3.4.1 Tips

1. **Correct sample preparation**
   - **Dilution:** Always start with serial dilutions of your sample to obtain a countable number of colonies (usually between 30–300 colonies per plate). Too many colonies on a plate can led to overcrowding and distort the results.
   - **Use a homogenizer or vortex:** Mix heterogeneous samples (*e.g.*, environmental swabs, soil samples) thoroughly to distribute the fungal cells evenly. This helps to avoid clumping or uneven distribution.

- **Sterility:** Use sterile equipment and aseptic techniques to avoid contamination by other microorganisms that could alter the CFU count.

2. **Selection of the medium**
   - **Agar type:** Select a suitable agar medium for fungal growth. Commonly used are YEPD, SDA, or PDA, depending on the fungal species.
   - **Selectivity:** If working with mixed microbial populations, choose a selective medium to inhibit bacterial growth (*e.g.* by using antibiotics or choosing a medium that favors fungi).
   - **pH considerations:** Ensure that the pH of the agar medium is appropriate for the fungal species. Fungi require a slightly acidic pH (around 5.5–6.5) for optimal growth.
3. **Plate inoculation**
   - For yeast-like fungi, use the plate method (spreading a small amount of the diluted sample on the surface of the plate). This avoids excessive fungal growth within the agar.
   - **Volume of inoculum:** Use a uniform volume (*e.g.* 100 μL) for each dilution. If you use higher volumes, distribute them evenly on the plate to ensure even distribution of the fungi.
4. **Incubation conditions**
   - **Temperature:** Incubate at the optimal temperature for fungal growth, usually 30 °C for most fungi.
   - **Time:** Depending on the strain of fungus, it may take 48–72 h for colonies to form. Be sure to incubate the plates appropriately and check colony formation regularly.
5. **Counting the colonies**
   - **Counting range:** Only count plates with between 30 and 300 colonies. Fewer colonies may not provide an accurate estimate of the fungal population, while too many colonies may lead to clumping or merging, resulting in miscounts.
   - **Consistency:** Count colonies in good lighting using a magnifying glass or colony counter to ensure clarity. Record the exact number of colonies or estimate if there are too many.
   - **Multiple plates per sample:** To increase accuracy, make several dilutions in duplicate or triplicate and average the results.

3.4.2 Troubleshooting

1. **Problem: No colonies formed**
   - **Possible causes**
     - The fungal strain used is overaged or decomposed.
     - Temperature and medium are not suitable for the fungus species.
     - If the sample is too dilute, there may not be enough fungal cells to form visible colonies.

- **Solutions**
  - Check the viability of the fungal strain by cultivating it on a different medium or under different incubation conditions.
  - Check the sample preparation process (*e.g.* make sure the sample has been properly homogenized).

2. **Problem: Too many colonies (overgrowth)**
   - **Possible causes**
     - Plating out too much sample or insufficient dilution can lead to overgrowth. This prevents an accurate CFU count as colonies merge or are overcrowded.
   - **Solutions**
     - Recalculate the DF to achieve a countable range of colonies.
     - Use smaller volumes of inoculum (*e.g.* 10–50 μL instead of 100 μL) or perform additional dilutions to ensure adequate spacing between colonies.
3. **Problem: Overlapping or merging colonies**
   - **Possible causes**
     - Overfilled plates can cause colonies to merge, resulting in inaccurate CFU counts.
     - Cell clumping can occur with some types of fungi.
   - **Solutions**
     - Mix the sample thoroughly before plating to prevent clumping.
     - Further dilute the sample so that the colonies are well dispersed.
     - Use a colony counter or magnifying glass for a more accurate count if overlap exists.
4. **Problem: Difficulty in colony identification (especially if they are contaminated)**
   - **Possible causes**
     - Bacterial contamination or other fungal species could be obscuring the colonies.
     - Some species or strains of fungi can form colonies with a similar appearance, making them difficult to distinguish.
   - **Solutions**
     - If you suspect contamination, add antibiotics (*e.g.* penicillin/streptomycin) to the medium to inhibit bacterial growth.
     - Use selective media or different agar types to differentiate fungal species.
     - Use selective staining techniques (*e.g.* methylene blue) or molecular methods for more accurate identification.
5. **Problem: Inconsistent results (between replicates or between different samples)**
   - **Possible causes**
     - Insufficient mixing of samples or uneven inoculation can lead to inconsistent results.
     - Some strains may exhibit slow growth or irregular colony formation, leading to variations in CFU count.

- **Solutions**
  - Ensure consistent handling and preparation of samples, including consistent dilution and smearing techniques.
  - If using strains with known variability, perform controls or replicate experiments for comparison.

6. **Problem: Low sensitivity when counting small colonies of slow-growing organisms**
   - **Possible causes**
     - Some fungal species grow slowly or form small colonies that are difficult to count manually.
   - **Solutions**
     - Incubate the plates for longer, sometimes up to 3–4 days for slow-growing fungi.
     - Use magnification devices or colony counters with higher sensitivity to detect small or slow-growing colonies.

Growth Curve Preparation for the Calculation of Fungal Cell Count Over Time

4

Abstract

Establishing a growth curve to calculate fungal cell count is a systematic approach in which a fungal culture is inoculated into a fresh medium and its growth is monitored over a period of time. This process begins with the selection of a suitable fungal strain and the preparation of a liquid medium that provides optimal growth conditions (Meletiadis et al, J Clin Microbiol 39:478–484, 2001). After inoculation, samples are taken at regular intervals, *e.g.* every few hours or days, depending on the expected growth rate of the fungus. Once the samples are collected, they are serially diluted to ensure that the resulting colonies are countable when plated on agar. This step is crucial as it allows for a more accurate determination of colony forming units (CFU/mL). After plating, the samples are incubated under controlled conditions and the colonies are counted as soon as they are visible. The CFU/mL is then calculated taking into account the dilution factor (DF) and the volume of the plated samples. To analyze the growth dynamics, the collected data is logarithmically transformed and displayed graphically. This helps to visualize the different growth phases: the lag, exponential, stationary and dying phases. Each of these phases provides valuable insights into the growth behavior of the fungus. For example, the delayed lag phase indicates adaptation to new conditions, while the exponential phase reflects rapid cell division. Statistical analysis allows you to evaluate differences in growth under different experimental conditions, such as changes in temperature, pH or nutrient availability. This rigorous method provides an accurate and comprehensive understanding of fungal growth dynamics over time, allowing researchers to draw meaningful conclusions about the physiology and ecology of fungi, as well as potential applications in biotechnology or medicine. Ultimately, growth curve analysis is an important tool in mycology that facilitates the study of fungal behavior in different environments (Salari, Salari, Electron Physician 9:3592–3597, 2017).

R. Pasrija, D. Kumari, *Protocols for Studying Pathogenic Fungi*, https://doi.org/10.1007/978-981-96-7047-5_4

Keywords

Colony-forming unit · Fungal culture · Dilution · Growth phases · Temperature · Mycology

Introduction This chapter describes the protocols from culture preparation to the generation of growth curves. The following topics are covered as follows:

4.1. Materials Required
4.2. Preparation of Media and Buffers
4.3. Protocol
4.4. Tips and Troubleshooting

4.1 Materials Required

4.1.1 Fungal Culture and Growth Media

- **Fungal strains**: Pure culture of the fungus species to be examined.
- **Growth medium:** Specific for the fungus being tested (*e.g.* yeast extract peptone dextrose (YEPD) broth or Roswell Park Memorial Institute medium (RPMI)-1640).
- **Sterile water or saline solution:** For dilutions or spore suspensions.

4.1.2 Culture Containers

- **Erlenmeyer flasks or culture tubes**: For growing the fungal culture.
- **Sterile Petri dishes:** If smears are required for CFU counting.

4.1.3 Sampling Tools

- **Pipettes and tips (sterile)**: Micropipettes with sterile tips or serological pipettes for accurate sampling.
- **Sampling vials:** Sterile vials/glass tubes for storing aliquots of the culture.

4.1.4 Measurement Devices

- **Spectrophotometer**: For measuring OD at specific wavelengths (*e.g.* 600 nm for turbidity).
- **Hemocytometer**: For direct microscopic counting of fungal cells.
- **Microscopy accessories (light microscope):** For observing the fungal cells.

4.1.5 Sterile Filtration and Dilution Materials

- **Filter paper/filter**: For sterile filtration if required.
- **Dilution tubes:** Sterile tubes for serial dilutions.
- **Sterile pipette tips:** To ensure contamination-free dilutions.

4.1.6 Time and Temperature Control

- **Incubator/shaker**: To maintain optimal growth conditions
- **Timer or stopwatch**: For uniform sampling intervals.

4.1.7 Sterilization and Safety Materials

- **Autoclave**: For sterilizing media and equipment.
- **Bunsen burner or alcohol lamp**: For aseptic techniques during sampling.
- **Gloves, lab coat and eye protection**: To ensure personal safety.

4.1.8 Data Recording and Analysis

- **Notebook or spreadsheet software**: For recording OD, CFU counts, or direct microscopic counts over time.
- **Graphing software (*e.g.* Excel, R, Python)**: For plotting the growth curve.

4.2 Preparation of the Media and Buffers

4.2.1 Preparation of YEPD Media (See Chap. 2)

4.2.2 Preparation of RPMI-1640 Media (See Chap. 2)

4.2.3 Preparation of PBS (See Chap. 3)

4.2.4 Preparation of YNB Medium and Agar

- **Materials required:** YNB powder, glucose, pH meter or pH strip, HCl or NaOH, agar, autoclave, *etc.*
- **Calculate the reagents:** For 1 L of liquid YNB media, mix 6.7 g YNB powder and 20 g of glucose. Adjust the quantities according to your specific requirements.
- **Dissolve the components:** In a sterile flask, add the appropriate amount of YNB powder and glucose to distilled water. Mix thoroughly until all solids have dissolved.

- **Adjust the pH value:** Use a pH meter or pH strips to check the pH of the solution. Adjust the pH to about 5.5–6.5 with HCl or NaOH if necessary.
- **Sterilize:** Autoclave the media at 121 °C for 15–20 min. Allow the media to cool to room temperature.
- **Storage:** If you do not use the liquid YNB media immediately, you can store them under sterile conditions at 4 °C for up to several weeks.

Preparation of YNB Agar Plates

- **Prepare YNB liquid media:** Follow the steps above to prepare 1 L of YNB liquid media.
- **Add agar:** For solid media, add 15 g of agar to the YNB liquid media before autoclaving. The ratio may vary depending on the desired solidity of the plates.
- **Sterilize:** Autoclave the mixture (YNB + agar) at 121 °C for 15–20 min.
- **Cooling and pouring the plates:** Allow the autoclaved medium to cool to around 50–55 °C (this prevents condensation). Pour the cooled agar solution into sterile Petri dishes (approximately 20–25 mL per dish).
- **Solidification:** Allow the agar plates to solidify for 15–30 min at room temperature.
- **Storage:** After solidification, store the YNB agar plates upside down (to avoid moisture accumulation) at 4 °C for up to 2–4 weeks.

Notes

- Ensure that all equipment and the work area are sterile to avoid contamination.
- Use appropriate aseptic techniques when inoculating and handling fungal cultures.
- If you require specific nutrients or supplements, you can add these depending on the fungal strain you are working with.

4.3 Protocol

Materials Required Fungal strain, sterile pipettes or micropipettes, sterile 15 mL or 50 mL tubes, YNB or other suitable medium, spectrophotometer, agar plates (YNB or selective medium), PBS, incubator, statistical software for growth curve analysis, *etc*.

4.3.1 Preparation of the Primary Culture and Application of the (Spotting) Cells to the Plates for Counting the CFU

1. Inoculate a single fungal colony in YEPD and incubate overnight at 30 °C with shaking.
2. After overnight growth, measure the culture's OD_{600} of the culture using a spectrophotometer and dilute the culture with 50 mL YNB in a sterile 50 mL tube to an OD_{600} of 0.1. Record this 0.1 OD as a 0-h reading and place the tube in the incubator at 30 °C.
3. Collect 100 μL of the culture at regular intervals (*e.g.* every 2 or 4 h) for CFU analysis the usual time points are 0, 2, 4, 6, 8, 10, 12 and 24 h.

4.3.2 Serial Dilution and Plating for CFU Counting

1. Prepare serial dilutions for each time point. For higher time points, dilute the serial dilutions to 10^{-10} to 10^{-15} to ensure countable colonies.
2. Label five sterile microcentrifuge tubes with 10^{-1}, 10^{-2}, 10^{-3}, 10^{-4}, and 10^{-5}, *etc.*
3. Add 900 μL of sterile saline to each tube.
4. Mix 100 μL of the cell suspension with 900 μL of sterile PBS in the first microcentrifuge tube. This results in a 1:10 dilution of the cells.
5. Take 100 μL of the cell suspension from the first tube (10^{-1}) and add it to the second tube (10^{-2}). Mix thoroughly.
6. Repeat the procedure to prepare the subsequent dilutions. For example, take 100 μL from tube 10^{-2} and add it to tube 10^{-3} and so on until the last dilution is made. Finally, discard the 100 μL of cells from the last tube (as shown in Fig. 4.1).
7. Add 5 μL of each dilution to the prepared YNB agar plates and incubate the plates at 30 °C for 48–72 h (48 h for *Candida*/*Saccharomyces* sp. and 72 h for *Cryptococcus* sp).
8. After incubation, record the number of colonies for each dilution.

4.3.3 Calculation of CFU/mL

1. Use the following formula to calculate the CFU/mL

$$\text{CFU} / \text{mL} = \text{number of colonies} \times \text{dilution factor}\left(\text{DF}\right) / \text{plating volume}\left(\text{in mL}\right)$$

Calculation example: If you get 2 colonies in 10^5 DF.

5 μL cells have = 2 colonies, then 1000 μL cells have = 2/5 × 1000 × DF (10^5) = 4 × 10^7 CFU/mL. The logarithm 10 of this CFU value would then be 7.602.

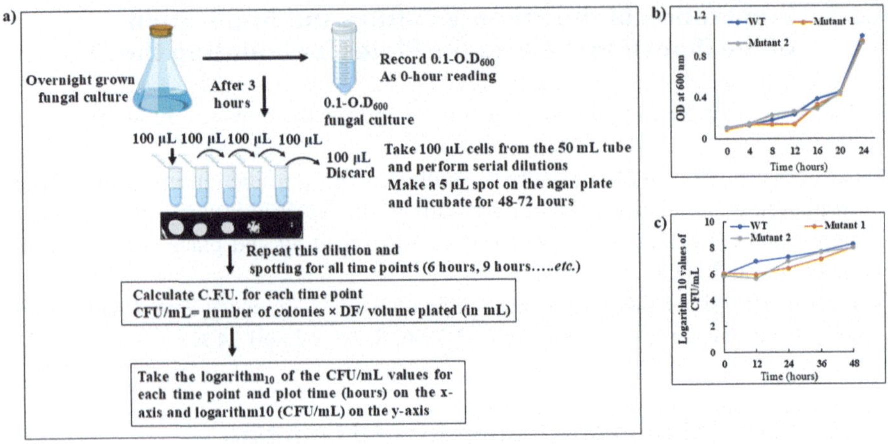

Fig. 4.1 (**a**) Flow chart showing the growth curve test for the fungal cell count over time. (**b**) A graph showing absorbance values on the y-axis and time on the x-axis. (**c**) A graph showing the logarithmic value (log base 10) for CFU/mL on the y-axis and time on the x-axis

4.3.4 Plotting of the Growth Curve

1. Take the logarithm 10 of the CFU/mL values for each time point.
2. Plot the time (hours) on the x-axis and the logarithm 10 (CFU/mL) on the y-axis.
3. The resulting graph illustrates the different phases of fungal growth: lag phase, logarithmic phase (exponential growth), stationary phase, and dying phase.

Note You can create a graph by plotting the OD_{600} values against time, or you can plot the logarithm (base 10) of the colony-forming units (logarithm 10 CFU) against time.

4.3.5 Statistical Evaluation

1. Make sure that the experiment is performed with biological replicates (*e.g.* at least 3 independent experiments).
2. Use software such as GraphPad Prism or Excel to analyze the data.
3. Calculate the mean and standard deviation for each time point.

4.4 Tips and Troubleshooting

4.4.1 Tips

1. **Selection of fungal strain and medium**
   - **Select a healthy fungal strain:** Use a freshly subcultivated strain with high viability. The strain should be adapted to the growth conditions (*e.g.* temperature, medium).
   - **Appropriate growth medium:** Use a suitable liquid growth medium for your fungal strain (*e.g.* SDB, YEPD, YNB or RPMI-1640). If you are working with dimorphic fungi, make sure you use a medium that supports both yeast and hyphae forms, if applicable.
   - **Sterility of medium:** Make sure all media are sterile to avoid contamination. Contamination can lead to inconsistent growth rates and affect the overall growth curve.
2. **Preparation of the inoculum**
   - **Inoculum density:** Start with a known inoculum density (typically about 1×10^6 cells/mL) to perform accurate CFU calculations.
     - For yeast, ensure that the culture is in the logarithmic phase of growth to avoid under- or overestimating the growth rate.
     - If your fungal species has a slow growth rate, start with a higher inoculum density to avoid longer lag phases.
   - **Homogenization:** Ensure that the inoculum is well mixed to avoid clumping of cells. For uniform mixing, you can use a vortex or homogenizer.
3. **Time points for sampling**
   - **Select suitable time points:** Choose multiple time points (*e.g.* 0, 2, 4, 6, 8, 12, 24, 48 h) based on the expected growth rate of your fungus.
   - **Sampling frequency:** For fast-growing fungi, sampling every 2–4 h during the exponential phase may be sufficient. For slower-growing fungi, you should increase the time interval.
4. **Methods for measuring growth**
   - **OD_{600}**: Use a spectrophotometer to measure the OD_{600}. This is a quick method to estimate growth, but it reflects the total biomass and not the number of viable cells. Always blank the spectrophotometer reading with the medium (without fungus) to correct for any absorption of the medium. The OD_{600} is useful during the logarithmic phase of growth as this is when it correlates most accurately with cell density.
   - **Direct cell count (CFU/mL)**: You can use serial dilution plating for each time point to determine the number of viable cells in terms of CFU/mL.
     - Use a hemocytometer or automated cell counter to estimate the cell density of liquid cultures, but this method works best with fungi that disperse well in the medium (do not clump).
5. **Incubation conditions**
   - **Keep the temperature constant:** Ensure a constant temperature throughout the experiment (*e.g.* 30 °C for most fungi)

- **Aeration:** If using liquid cultures, ensure adequate aeration by shaking at an appropriate speed (*e.g.* 150–200 rpm) to avoid settling or clumping of the fungal cells.

6. **Replicates**
   - **Use biological replicates:** Run the growth curve in duplicate or triplicate to ensure reproducible results.
   - **Technical replicates:** Take multiple samples at each time point to reduce variability due to sample handling.

4.4.2 Troubleshooting

1. **Problem: No growth or slow growth of fungal culture in YNB medium**
   - **Possible causes**
     - **Suboptimal inoculum**: The original fungal inoculum may be too weak or not actively growing.
     - **Wrong medium**: The YNB medium may not be suitable for the specific fungal strain.
     - **Temperature problems**: The culture may not have been incubated at the correct temperature (30 °C).
   - **Solutions**
     - Check the growth requirements of the fungal strain (temperature, medium). Some strains may require special nutrients or additional supplements in the medium.
     - Use a fresh colony from the YEPD plate for inoculation to ensure an active culture.
     - Re-check the temperature of the incubator to ensure it is set to 30 °C.
     - Ensure that the YNB medium is properly prepared and sterilized (autoclave before use).
2. **Problem: The CFU numbers are too high or too low**
   - **Possible causes**
     - **Over/under dilution:** An incorrect DF leads to too many or too few colonies on the plate.
     - **Incorrect plating volume**: You have plated too much or too little (an incorrect volume can distort the CFU calculations).
   - **Solutions**
     - If the CFU counts are too high, dilute the sample further (try dilutions greater than 10^{6}, for example).
     - If the CFU counts are too low, reduce the dilution.
     - Make sure you are plating out the correct volume (usually 5 µL).
     - Dilute the sample up to 10^{10} to 10^{15} and discard the previous dilution tubes. Illuminate only the last 5–6 dilution tubes to ensure that the colonies remain countable at higher time points. This procedure helps to avoid overcrowding and ensures accurate colony count.

3. **Problem: The colonies merge on the plate**
   - **Possible causes**
     - **Too many cells were plated**: Over-concentration of cells leading to colony merging.
     - **Incorrect dilution preparation**: Inaccurate serial dilution leads to too many cells being plated.
   - **Solutions**
     - Use more diluted samples, especially for higher time points. For example, increase the DF for late time points.
     - Make sure to vortex each dilution tube properly before plating to ensure even distribution of cells.
     - When spotting, place a template with spots under the plate to avoid the spots being too close together.
4. **Problem: The growth curve shows irregular or unexpected results**
   - **Possible causes**
     - **Incorrect CFU/mL calculation**: Misinterpretation of the colony count or DF can lead to incorrect CFU calculations.
     - **Inconsistent sampling**: Sampling at irregular intervals or inconsistent preparation of samples at all times.
   - **Solutions**
     - Check the CFU/mL calculation formula and ensure that the correct DF and plating volume are used: CFU/mL = number of colonies × DF/plating volume (in mL)
     - Ensure uniform sampling and dilution at each time point. Regular intervals (*e.g.* every 2 h) help to monitor growth accurately.
     - Provide biological replicates (at least 3 independent samples) to minimize experimental variability.
5. **Problem: The analysis of the growth curve in the graphing software is incorrect**
   - **Possible causes**
     - **Error in logarithmic transformation**: Error in converting CFU/mL values to logarithm10 (especially with very low or high CFU numbers).
     - **Error during data input**: Input error when entering data into the software or use of incorrect time points.
   - **Solutions**
     - Double check thelogarithm10 transformation. Make sure the CFU/mL values are within an appropriate range for the plot.
     - Use the correct software settings and carefully enter the data into statistical software such as GraphPad Prism or Excel.
     - Make sure the time points are consistent and labelled correctly to create an accurate graph.

Microtiter Assay for Drug Susceptibility Testing of Pathogenic Fungi in Broth

5

Abstract

Microtiter assay is a widely used method for drug susceptibility evaluation of pathogenic fungi in broth and provides a standardized, reproducible and efficient approach for fungal susceptibility testing. This method is important for the determination of the minimum inhibitory concentration (MIC) of antifungal drugs and helps in the development of treatment strategies against fungal infections. The assay is based on exposing fungal cells to different concentrations of antifungal agents in a liquid medium and monitoring their growth (Berkow et al, Clin Microbiol Rev 2020). In a typical microtiter assay, fungal spores or cells are inoculated into wells of a microtiter plate containing serial dilutions of the antifungal agent in a suitable growth medium, such as RPMI-1640. The plates are incubated at specific optimal conditions, usually between 25 and 35 °C, for 24–72 h, depending on the fungal species. Growth is assessed either visually by observing turbidity or quantitatively by spectrophotometric measurements at specific wavelengths. The MIC is determined as the lowest concentration of the antifungal agent that inhibits visible fungal growth or reduces metabolic activity (EUCAST, Clin Microbiol Infect 14(4):398–405, 2008). The scalability of the microtiter test, the minimal reagent requirement and the adaptability to high-throughput formats make it the preferred choice for susceptibility testing of clinically important fungi such as *Candida* sp. and *Aspergillus* sp. The ability to check eight drugs/agents in separate rows/series, the scalability of the microtiter test, the minimal reagent requirement and the adaptability to high-throughput formats make it the preferred choice for susceptibility testing of clinically important fungi (CLSI, Reference method for broth dilution antifungal susceptibility testing of yeasts; approved standard-third edition, CLSI Document M27-A3. Clinical and Laboratory Standards Institute, Wayne, 2008). Despite its advantages, the test also has its limitations. These include potential variability in inoculum preparation, drug solubility issues, and the challenge of testing slow-growing or biofilm-forming fungi.

R. Pasrija, D. Kumari, *Protocols for Studying Pathogenic Fungi*,
https://doi.org/10.1007/978-981-96-7047-5_5

Nevertheless, the microtiter assay remains an indispensable tool for the evaluation of antifungal drugs and contributes to the understanding of fungal resistance mechanisms and the optimization of therapeutic regimens.

Keywords

Microdilution · Fungi · Microtiter · CLSI · Susceptibility · Yeast

Introduction This chapter describes the protocols from the stock preparations to the microtiter assay for analyzing the MIC of any test material against yeasts. The following topics are covered as follows:

5.1. Materials Required
5.2. Preparations
5.3. Procedure
5.4. Tips and Troubleshooting

5.1 Materials Required

Microtiter plates (usually 96-well plates), fungal inoculum (spores or mycelium of the fungal pathogen), RPMI1640/yeast extract peptone dextrose (YEPD)/potato dextrose agar (PDA) or another suitable growth medium (depending on the pathogen), test substances (*e.g.*, antifungals), sterile distilled water or buffer, pipettes and pipette tips (preferably multichannel pipettes for efficiency), sterile tools (*e.g.* microcentrifuge tubes, micropipettes, sterile glassware, *etc.*), incubator (set to the appropriate temperature for fungal growth) and plate reader (for absorbance, fluorescence, or luminescence).

5.2 Preparations

5.2.1 Preparation of Media

- Yeast extract peptone dextrose media (YEPD) (see Chap. 2)
- Preparation of RPMI-1640 media (see Chap. 2)

5.2.2 Preparation of Drug Stocks

- **Solvent Selection:** Select a suitable solvent for dissolving the antifungal drug based on its solubility properties:
  - Dimethyl sulfoxide (DMSO): Commonly used for hydrophobic drugs such as azoles and polyenes.
  - Sterile water: Suitable for water-soluble drugs.
  - Ethanol or methanol: Only used when DMSO or water are not suitable.

- **Preparation of stock solution [1]**
  - **Determine the concentration:** Prepare a highly concentrated stock solution (*e.g.* 1000–10,000 μg/mL) for serial dilution during the test. The concentration should ensure accuracy and minimize the volume of solvent in the final assay.
  - **Weighing the drug:** Accurately weigh the required amount in a sterile tube using an analytical balance. Dissolve the drug in the selected solvent under sterile conditions. If necessary, shake the solution to dissolve it completely.
  - **Sterilize:** Filter the prepared stock solution through a 0.22 μm syringe filter to ensure sterility and remove particulate impurities. This step is particularly important when working with clinical isolates or preparing long-term stock solutions.
  - **Aliquoting and storage:** Divide the stock solution into small aliquots for single use to avoid repeated freezing and thawing cycles that can degrade the drug. Then store the aliquots under the recommended conditions, usually at −20 °C or − 80 °C, depending on the stability of the antifungal. Label each aliquot with the name of the drug, concentration, solvent, date of manufacture and storage conditions.
  - **Documentation:** Keep detailed records of the preparation of the supplies, including the source of the drug, batch, lot number, weight measured, solvent used, final concentration, and storage conditions. This documentation ensures reproducibility and traceability.

5.2.3 Preparation of Fungal Culture

There are slight variations in the preparation of fresh yeast and mold cultures for microtiter assays that must be taken into account. The preparation of yeast cultures is described in Chap. 3, while the protocol for molds such as *Fusarium* sp. is explained in more detail below:

- Streak or spot-inoculate the fungal isolate onto PDA.
- Incubate the plate at 25–30 °C for 5–7 days until sporulation is observed. Check under the microscope for the presence of conidia to confirm adequate growth.
- Add 10 mL of sterile saline solution containing 0.05% Tween 80 (NST) to the surface of the fungal culture.
- Gently rub the surface with a sterile loop or spreader to disperse the conidia into a sterile tube.
- Filter the suspension through sterile cotton filters to remove hyphal fragments and obtain a pure spore suspension (without filaments) for the microtiter assay.

5.3 Procedure

1. Add 100 μL of the sterile medium (*i.e.*, generally RPMI1640/YEPD for yeast and PDA for molds) for fungi to each well of a flat-bottomed 96-well microtiter plate.
2. Row 11 is the growth control well, which is treated as a negative control and only requires the addition of medium (200 μL).
3. Row 12 of the microdilution plate is used as a sterility control and contains only a drug-free medium.
4. Add the required drugs/test chemicals from the prepared stocks in serial dilution from the first to the tenth well so that the final volume in each well is 100 μL.
5. Then adjust the concentration of the fungal culture in a spectrophotometer to 0.1 OD (0.1 OD corresponds to 1×10^6 cells/mL of yeast cells); or (0.7 on the McFarland scale in the case of yeast)/0.5 McF for molds, by adding sterile 0.9% saline. If you are working with yeast cultures, then dilute the inoculum to 1:1000 with sterile medium to achieve a final concentration of ~0.5–2.5 × 10^3 cells/mL. For molds such as *Fusarium* sp., dilute the adjusted 0.1 OD inoculum to 1:50 to achieve ~1×10^4 to 5×10^4 CFU/mL.
6. Add 100 μL of the diluted inoculum suspension to the first 10 wells and the 12th well (positive control) and do not add culture to the 11th well.
7. After a 48–72-h incubation at 28–35 °C (*i.e.* for *Candida* sp. 48 h at 30 °C, *Cryptococcus* sp. 72 h at 30 °C, *Fusarium* sp. 72–96 h at 28 °C and *Aspergillus* sp. 48 h at 35 °C), examine the microdilution plate for fungal growth using a microplate reader at 595 nm or visually using a concave plate reader.
8. The MIC value is the lowest concentration of the antifungal agent that prevents fungal growth (as shown in Fig. 5.1).
9. Quality control (QC) organisms, such as *Candida krusei* ATCC 6258 and *Candida parapsilosis* ATCC 22019, should be included in each experiment and tested identically. If the MICs of the QC strains are within the specified ranges, the reliability of the overall data is increased. The MIC ranges (μg/ml) of these strains for amphotericin B are:
   - *C. krusei* ATCC 6258: 0.5–2
   - *C. parapsilosis* ATCC 22019: 0.25–2

Notes:

- Ensure that all reagents and fungal cultures are freshly prepared and sterile.
- Note the volume of the final test mixture (inoculum + medium + drug) and ensure that it is the same in all wells.
- Perform the test in triplicate or multiples to ensure reproducibility and accuracy.
- It is important to optimize the incubation time for the fungal pathogen to be tested as growth rates may vary.

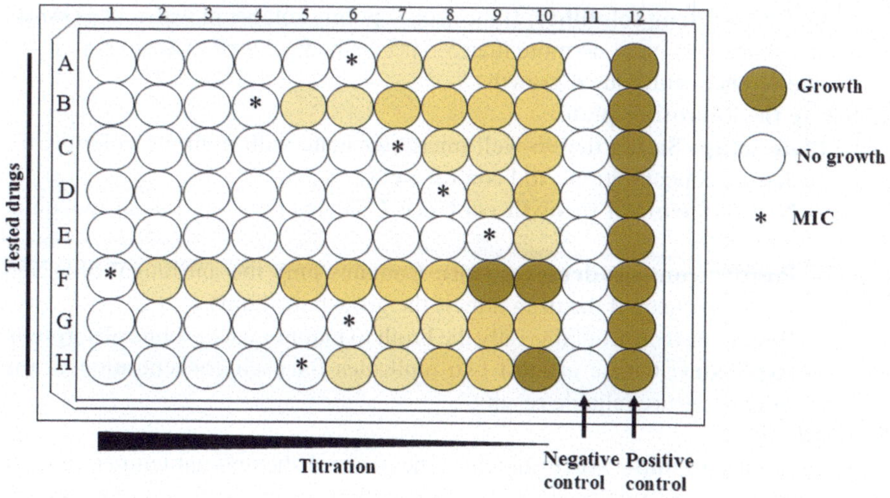

Fig. 5.1 Illustration of the microtiter test, where the (*) mark indicates the MIC endpoint in the different wells

5.4 Tips and Troubleshooting

5.4.1 Tips

1. **Selection and preparation of the fungal strain:**
   - **Select a representative fungal strain:** Choose the fungal species or strain that is clinically relevant to your study. Use a freshly cultured strain (from an agar plate) in the logarithmic growth phase to ensure active cell division.
   - **Inoculum preparation:** Standardize the inoculum density with a spectrophotometer or hemocytometer to achieve a specific cell concentration (typically 1×10^6 cells/mL). This ensures reproducibility and accurate interpretation of the results. If using an inoculum suspension, ensure that it is well mixed to avoid clumping of fungal cells.
2. **Selection of the medium:**
   - **Suitable broth medium:** Use a suitable liquid medium for fungal growth. Ensure that the medium supports the optimal growth of your fungal strain and does not interfere with the activity of the antifungal agent.
   - **pH considerations:** The pH of the medium should be adjusted (usually to 7.0 or slightly acidic) to prevent interference with antifungal activity.
3. **Preparation of the drug:**
   - **Antifungal stock solutions:** Prepare antifungal drug solutions at high concentrations and serially dilute to obtain a range of drug concentrations in the microtiter plate. For fungicides, the working concentration range is usually between 0.001 and 64 μg/mL. Ensure that the active ingredients are completely dissolved and avoid using expired or improperly stored stock solutions.

- **DMSO for drug solubility:** If the drug is poorly soluble in water, use DMSO at a concentration of no more than 5–10% in the final well to avoid toxicity or interference with fungal growth.

4. **Set up the microtiter plate:**
   - **Plate setup:** Set up the 96-well microtiter plate with multiple columns for each drug concentration. Add control wells:
     - **Negative control (growth control):** Contains only the fungal inoculum and broth.
     - **Positive control (drug control):** Contains only the antifungal drug and broth (to check if the drug affects the growth medium).
     - **Blank control:** Contains only the broth to monitor background absorption.
     - **Replicates:** Create at least two replicates for each concentration of the drug to ensure reliable results.
5. **Incubation:**
   - **Incubation temperature:** Incubate the plates at the optimal temperature for fungal growth (usually 30 °C for most pathogens or 37 °C for some dimorphic fungi such as *C. albicans*).
   - **Incubation time:** Incubate the plates for 24–48 h, depending on the fungal strain and growth rate. A longer incubation time may be required for slow-growing fungi.
6. **Read the assay (endpoint determination):**
   - **Visual observation:** visually inspect the wells for fungal growth after incubation. Growth is indicated by turbidity or the formation of a visible fungal mass. The MIC is the lowest concentration of the drug/active ingredient at which no visible growth is observed.
   - **Alternative endpoint measurement:** You can also measure fungal growth using a microplate reader at 595 nm. The lowest drug concentration at which the growth is minimal is the MIC endpoint.
   - **Confirm growth:** To confirm growth or inhibition, you can subcultures from wells without growth on solid agar to ensure no viable colonies are present.

5.4.2 Troubleshooting

1. **Problem: Poor or inconsistent fungal growth**
   - **Possible causes:**
     - Incorrect composition of the medium or incorrect pH value.
     - Insufficient concentration of inoculum (too high or too low).
     - Suboptimal incubation conditions (temperature, humidity, or aeration).
     - Contamination with bacteria or other fungi.
   - **Solutions:**
     - **Check the preparation of the medium:** make sure that the growth medium is correctly formulated and sterile.
     - **Check the concentration of the inoculum:** Standardize the fungal spore/mycelium suspension with a hemocytometer or spectrophotometer.

- **Optimize the incubation conditions:** Ensure that the temperature, light and aeration conditions meet the requirements of the fungal species.
- **Control contamination:** Work in a sterile environment and use antibiotics if necessary (in case of bacterial contamination).

2. **Problem: Strong background noise during absorption**
   - **Possible causes:**
     - Residual media components interfering with the assay.
     - Optical artifacts (*e.g.* bubbles or debris in the wells).
   - **Solutions:**
     - **Include controls:** Include blank wells (medium without fungi) and negative controls to subtract background noise.
     - **Avoid air bubbles:** Gently tap or centrifuge the plate to remove air bubbles before reading.
     - **Check for contamination:** Ensure the test reagents are fresh and not contaminated.
3. **Problem: Uneven growth or signals in the wells**
   - **Possible causes:**
     - Poor mixing of the inoculum or reagents.
     - Edge effect due to evaporation or temperature gradients.
     - Inconsistent pipetting or handling.
   - **Solutions:**
     - **Ensure uniform mixing:** Mix the fungal suspension and reagents thoroughly before adding them to the wells.
     - **Minimize edge effects:** Fill the outer wells with sterile water or buffer and use only the inner wells for the assay.
     - **Standardize pipetting:** Use calibrated micropipettes and ensure a consistent technique.
4. **Problem: Contamination in the wells**
   - **Possible causes:**
     - Poor aseptic technique.
     - Contaminated stock solutions or fungal inoculum.
     - Contamination from the air during plate handling.
   - **Solutions:**
     - Improve aseptic technique: Work under a laminar flow hood and sterilize your tools thoroughly.
     - **Check the purity of the material:** Examine the fungal inoculum regularly under a microscope for contamination.
     - Use suitable closures: Cover plates with sterile lids or breathable seals/closures to reduce airborne contamination.
5. **Problem: Problems with the reproducibility of the data**
   - **Possible causes:**
     - Variability in the preparation of the inoculum
     - Inconsistent incubation conditions.
     - The multichannel pipette may dispense different volumes into different tips.
     - Operator error during handling or measurement.

- **Solutions:**
  - **Standardize protocols**: Prepare inoculum and reagents the same way every time.
  - The multichannel pipette should be handled carefully to ensure an equal volume in each tip.
  - **Use replicates:** Add biological and technical replicates to account for variability.
  - **Automate processes**: Use automated pipetting systems and plate readers where possible to avoid manual errors.

References

1. Lovero G, Giglio OD, Rutigliano S, Diella G, Caggiano G, Montagna MT (2017) In vitro antifungal susceptibilities of *Candida* species to liposomal amphotericin B, determined using CLSI broth microdilution, and amphotericin B deoxycholate, measured using the Etest. J Med Microbiol 66(2):213–216. https://doi.org/10.1099/jmm.0.000402

Disk Diffusion Assay for Testing the Drug Sensitivity of Pathogenic Fungi in Agar

6

Abstract

The disk diffusion method is one of the oldest and most widely used techniques for sensitivity testing, especially in clinical laboratories working in a resource constrained environment. This method is not only cost-effective, but also provides consistent results that are easy to interpret, making it a valuable tool in antimicrobial resistance studies (Berkow et al, Clin Microbiol Rev 33:e00069-19, 2020). In this method, paper disks, each infused with a specific concentration of an antifungal drug, are placed on the surface of an agar plate that has been inoculated with the fungal isolate. Once the disks are attached, they are incubated under optimal conditions to allow fungal growth and diffusion of the drug. After incubation, the size of the clear area around each disk, known as the zone of inhibition, is measured. This zone indicates the extent of drug diffusion through the agar and reflects the isolate's susceptibility to the antifungal agent tested. A larger zone usually indicates higher susceptibility, while a smaller or absent zone indicates resistance. The Clinical and Laboratory Standards Institute (CLSI) has established standardized protocols for the disk diffusion assay, particularly for yeast susceptibility testing (2nd edition CLSI document M44-A2, 2009). This ensures the reliability and reproducibility of results in different laboratories. In addition, the disk diffusion method enables the simultaneous testing of several antifungal agents, providing a comprehensive understanding of the resistance profile of an isolate. In addition to its practical application in the clinical field, the disk diffusion method is also an indispensable tool for researchers working on fungal pathogens. As this method provides a clear visual representation of susceptibility, it helps to monitor emerging resistance patterns and ultimately contributes to improved patient care and treatment strategies for fungal infections (Sanchez et al, Microbiol Res 252:126867. 2021). Overall, the disk diffusion method remains a cornerstone of antifungal susceptibility testing for fungi and bridges the gap between research and clinical application.

R. Pasrija, D. Kumari, *Protocols for Studying Pathogenic Fungi*,
https://doi.org/10.1007/978-981-96-7047-5_6

Keywords

Disk diffusion · Susceptibility · Yeast · Agar plate · CLSI

Introduction This chapter describes a detailed protocol for determining the effective drug concentration against yeasts on agar plates, from the preparation of the starting material to the disk diffusion test. Topics covered include:

6.1 Materials Required
6.2 Preparations
6.3 Procedure
6.4 Tips and Troubleshooting

6.1 Materials Required

The following materials and equipment are generally required for the disk diffusion test to check the drug sensitivity of pathogenic fungi on agar:

6.1.1 Agar Plates

- Yeast Extract Peptone Dextrose (YEPD), Sabouraud Dextrose Agar (SDA) or Potato Dextrose Agar (PDA) are commonly used for the cultivation of fungi. These provide nutrients that promote fungal growth.
- The agar plates should be freshly prepared to ensure consistency.

6.1.2 Fungal Culture

- **Fungal Pathogen:** The fungal pathogen to be tested (*e.g.*, *Candida* sp., *Aspergillus* sp., *Cryptococcus* sp., *Fusarium* sp., *etc.*).

6.1.3 Antifungal Disks

- Commercially available antifungal disks (*e.g.*, Amphotericin B, Fluconazole, Itraconazole, Voriconazole, Ketoconazole, *etc.*).
- Ensure that the disks are stored according to the manufacturer's instructions (usually refrigerated) to maintain their effectiveness.

6.1.4 Aids for the Inoculum Preparation

- **Sterile Swabs or loops:** Sterile swabs or loops: These are re used to spread the fungal inoculum evenly on the agar surface.
- **Sterile saline solution or broth:** To prepare the inoculum suspension that meets the 0.1 OD/0.5 McFarland standard.

6.1.5 Incubation

- **Incubator:** Suitable for maintaining the desired temperature for fungal growth (typically 25–30 °C).
- **Incubation time:** Normally, 24–48 h, depending on the type of fungus being tested.

6.1.6 Sterile Forceps or Tweezers

- Used to handle the antimicrobial disks and place them on the agar surface without contaminating it.

6.1.7 Caliper or Ruler

- Used to measure the zones of inhibition to assess the efficacy of the antifungal agents.

6.1.8 Biosafety and Sterilization Equipment

- Autoclave: For sterilization of materials such as glassware and tools.
- Laminar flow hood: To prevent contamination during inoculation and disk placement.

6.1.9 Cleaning Agents

- Disinfectants: Such as 70% ethanol or other suitable disinfectants for cleaning surfaces, equipment and after handling fungal cultures.
- Gloves and lab coats for safety.

6.2 Preparations

6.2.1 Preparation of Culture Media and Agar (See Chap. 2)

6.2.2 Preparation of the Fungal Culture (See Chap. 3)

6.2.3 Preparation of the Paper Disks

- **Cutting the disks:**
  - Using a sterile hole punch or a sharp circle cutter, cut the filter paper into disks of uniform size, usually 6 mm in diameter.
  - If you are using pre-cut disks, make sure they are sterile.
- **Sterilize the disks:**
  - Place the cut paper disks in a sterile Petri dish or sterilizable container.
  - Make sure that the disks are arranged in a single layer without overlapping.
- **Autoclaving the paper disks**
  - **Prepare the autoclave:** Make sure the autoclave is clean and working properly. Make sure that the water tank in the autoclave is full and the chamber is free of foreign objects.
  - **Load the disks:** Place the Petri dish or the container with the paper disks in the autoclave. If using a container, ensure it is ventilated so that the steam can circulate.
  - **Set the autoclave parameters:** Start the autoclave with the default standard parameters, including temperature and pressure of 121 °C and 15 psi respectively, for 15–20 min.
  - **Removing the disks:** After the autoclave has cooled down, carefully remove the container or petri dish containing the sterilized disks from the autoclave using sterile forceps. Then allow the disks to cool down in a sterile environment.
  - **Notes**
    - Ensure that all materials and tools used are sterile to avoid contamination.
    - Perform all steps related to the handling of the disks in a sterile environment, *e.g.* in a laminar flow hood, to maintain their sterility.
    - Regularly check the calibration and functionality of the autoclave to ensure effective sterilization.

6.2.4 Preparation of Drug Stocks (See Chap. 5)

6.3 Procedure

- Adjust the fungal culture concentration with sterile 0.9% saline solution to an OD of 0.1 and achieve a final concentration of approximately 1×10^6–1×10^7 cells/mL. Alternatively, the McFarland scale can be used: a value of 0.7 is recommended for yeast, while 0.5 McFarland is suitable for filamentous fungi. Use a spore suspension for filamentous fungi and a yeast cell suspension for unicellular fungi.
- Dip a sterile cotton swab (or use a pipette) into the prepared culture and place the swab with the inoculum in the center of the agar plate and streak in a zig-zag pattern.
- Rotate the plate by 90° and repeat the swabbing procedure three times (see Fig. 6.1). This ensures an even distribution of the inoculum and confluent growth of culture after incubation.
- Allow the surface of the medium to dry for 3–5 min, but no longer than 15 min to allow excess moisture to be absorbed.
- Using sterile forceps or a disk dispenser, place the autoclaved filter paper disk on an agar plate and dispense the drug into the center of the disk.

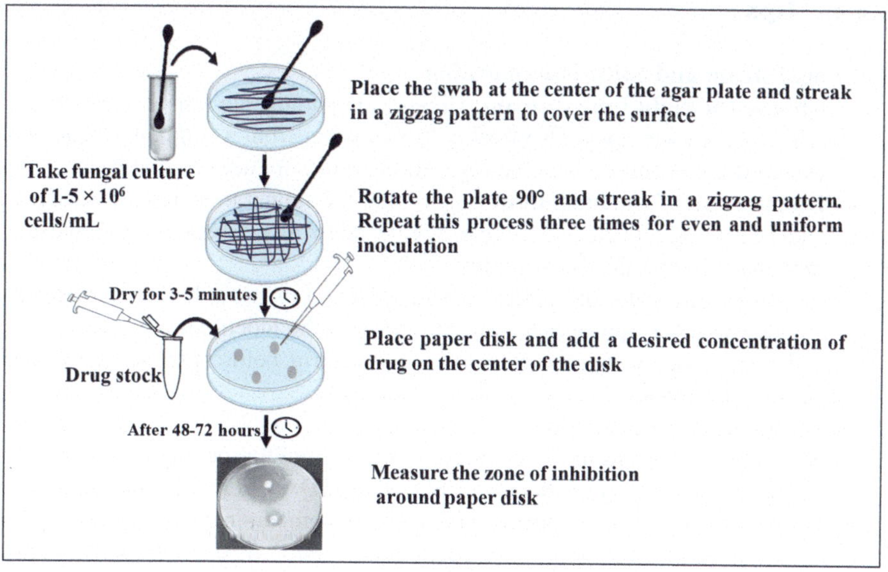

Fig. 6.1 Flow chart showing the disk diffusion test procedure for checking the antifungal activity of the test material

- Immediately press lightly with the instrument to ensure complete contact between the disk and the agar surface. Do not move the disk once it has come into contact with the agar surface, as some of the drug will diffuse immediately.
- Position the disks so that the minimum center-to-center distance is 24 mm and no closer than 10–15 mm from the edge of the Petri dish. A maximum of six disks can be placed in a 9 cm Petri dish and 12 disks on a 150 mm plate. Reduce the number of disks per plate if the inhibition zones overlap.
- Incubate the plates at 35 ± 2 °C for 48 h for yeasts (*e.g. Candida* sp), 72 h for *Cryptococcus* sp., and at 25–30 °C for 72 h for molds (*e.g. Fusarium* sp., *Aspergillus* sp., *etc.*). Some slow-growing molds (*e.g.* certain *Fusarium* sp) may require up to 96 h of incubation.
- Then measure and record the diameter of the inhibition zones using a ruler set at 0.5 mm. Round the zone diameter up to the nearest mm.
- Include quality control (QC) organisms such as *Candida albicans* ATCC 90028, *Candida krusei* ATCC 6258 and *Candida parapsilosis* ATCC 22019 in each experiment, following the same test protocols. The disk diffusion ranges for these QC strains against fluconazole are as follows: *C. albicans* ATCC 90028: 32–43 mm, *C. parapsilosis* ATCC 22019: 26–37 mm and *C. krusei* ATCC 6258: 6–17 mm [1].
- The step-by-step procedure and the corresponding result are shown in Fig. 6.1.

6.4 Tips and Troubleshooting

6.4.1 Tips

1. **Fungal strain and culture preparation**
   - **Choose the right fungal strain:** Choose the pathogenic fungal strain that is relevant for your study. Use freshly grown fungi from a solid agar plate and ensure that the culture is in the logarithmic phase for active growth.
   - **Preparation of the inoculum:** Ensure that the fungal inoculum is standardized to the correct cell density (*e.g.* 1×10^6 cells/mL). This ensures consistent inoculation and reduces variability.
     - You can measure the optical density (OD_{600}) or use a hemocytometer to determine the appropriate concentration for the inoculum.
     - Ensure that the fungal suspension is uniform before plating to prevent uneven growth.
2. **Selection of the agar medium**
   - **Suitable agar medium:** Select the correct medium for the fungal species you are testing. Common media are YEPD, SDA, PDA and RPMI 1640 agar.
   - **Preparation of the medium:** Make sure the medium is sterile and well-mixed to avoid problems with pH value or nutrient distribution. The pH should generally be around 7.0 to ensure optimal fungal growth.
   - **Agar depth:** Standardize agar depth to 4–6 mm, as deeper or shallower agar layers may affect diffusion rates and inhibition zone size.

3. **Drug and disk preparation**
   - **Antifungal agents:** Use freshly prepared drug solutions or commercially available impregnated disks. If you prepare your own disks:
     - Ensure that the drugs are dissolved in the correct solvent (*e.g.* DMSO, water, ethanol) without exceeding the maximum allowable concentration (usually ≤1–2% DMSO).
     - Use serial dilutions to impregnate the disks with a range of drug concentrations.
   - **Use the right disk size:** Ensure that the disks are the standard size (usually 6 mm diameter), as this will ensure consistent results and reliable comparison of inhibition zones.
4. **Inoculation and placement of the disk**
   - **Agar plate inoculation:** Use a sterile swab to inoculate the surface of the agar with the fungal suspension.
   - **Uniform inoculum:** Make sure that the inoculum is evenly distributed over the entire agar surface. Avoid clumping of fungal cells as this can lead to irregular growth patterns.
   - **Placement of disks:** Use sterile forceps to place the antifungal disks on the inoculated agar surface. Make sure that they are evenly distributed (*e.g.* at least 1–2 cm apart) to avoid interference between the inhibition zones.
5. **Incubation conditions**
   - **Temperature:** Incubate the plates at the optimum temperature for fungal growth. Normally, 30 °C is ideal for most fungal pathogens, but 37 °C may be required for certain species.
   - **Incubation time:** Incubate the plates for 48 h, depending on the growth rate of the fungal species. Slow-growing fungi (*e.g.* molds) may require longer incubation times of up to 96 h.
   - **Humidity:** Ensure adequate humidity in the incubator to prevent the agar plates from drying out during incubation.
6. **Measuring the inhibition zone**
   - **Accurate measurement:** Measure the diameter of the inhibition zone (clear area around the disk) using a caliper or ruler.
   - **Evaluation:** The size of the zone of inhibition is related to the efficacy of the drug. Compare the observed zone with the established breakpoint standards for the respective drug and fungal species.
     - A large zone indicates good activity of the drug, while a small or absent zone indicates resistance or an ineffective drug concentration.
   - **Record the Zone Size:** Note the zone diameter (in mm) for accurate analysis and comparison between different drugs or conditions.

6.4.2 Troubleshooting

1. **Problem: Uneven or no inhibition zone**

- **Possible causes:**
  - **Insufficient concentration of inoculum**: The fungal suspension is too dilute or too dense.
  - **Incorrect placement of the disk: The** disk not in full contact with the agar surface.
  - **Inactive antimicrobial agent**: The antimicrobial agent may have degraded.
  - **Agar too dry**: Impairs diffusion of the antimicrobial agent.
  - **Problems with the composition of the medium**: pH value or nutrient content are unsuitable for fungal growth.
- **Solutions:**
  - Standardize the density of the inoculum using a spectrophotometer or hemocytometer.
  - Ensure that the disks are gently and evenly pressed onto the agar.
  - Use fresh antimicrobials and store according to the manufacturer's recommendations.
  - Avoid over-drying the agar plates by keeping them under lock and key during preparation.
  - Prepare media according to standardized protocols and check pH prior to use.

2. **Problem: Irregular or asymmetric zones**
   - **Possible causes:**
     - **Uneven distribution of inoculum**: The fungal suspension is not evenly distributed on the agar.
     - **Disks placed too close together**: Overlapping diffusion zones.
     - **Irregularities on the agar surface**: Uneven pouring or bubbles in the agar.
     - **Uneven incubation conditions**: Temperature gradients or uneven humidity.
   - **Solutions**:
     - Use a sterile spreader to ensure even distribution of the fungal inoculum.
     - Place the disks at a sufficient distance to avoid overlapping.
     - Ensure that the agar is poured evenly and without bubbles; allow it to solidify on a flat surface.
     - Use a well-calibrated incubator with uniform temperature and humidity distribution.
3. **Problem: Small or weak inhibition zones**
   - **Possible causes:**
     - **Low antimicrobial potency**: Degraded or insufficient concentration of antimicrobial agent/active ingredient.
     - **Too thick agar**: Impairs diffusion of the antimicrobial agent.
     - **Fungal resistance**: The strain may be resistant to the antimicrobial agent tested.
   - **Solutions:**
     - Check the efficacy and concentration of the antimicrobial agent.
     - Maintain a uniform agar thickness (typically 4 mm for standard assays).

   - Confirm fungal susceptibility using alternative methods if necessary to determine resistance.
4. **Problem: Overlapping zones of inhibition**
   - **Possible causes:**
     - **Disks placed too close together**: This leads to overlapping diffusion zones.
     - **Too many disks on a single plate**: Leads to crowded zones.
   - **Solutions:**
     - Place the disks evenly and keep a sufficient distance between them (at least 24 mm).
     - Use larger plates or fewer disks per plate.
5. **Problem: Lack of reproducibility between experiments**
   - **Possible causes:**
     - **Variations in inoculum preparation**: Different fungal densities or viability.
     - **Variations from batch to batch in agar or disks**: Inconsistent quality of media or disk.
     - **Operator variability**: Inconsistencies in technique or execution.
   - **Solutions:**
     - Use standardized methods to prepare inoculum and verify density prior to use.
     - Use high-quality, consistent batches of media and antimicrobial disks.
     - Provide thorough training to ensure consistent methodology for all staff.
6. **Problem: Contamination issues**
   - **Possible causes:**
     - **Non-sterile environment**: Contamination during plate and disk preparation or inoculation.
     - **Cross-contamination between disks**: Improper handling of antimicrobial disks and plates.
   - **Solutions:**
     - Work in a sterile environment (*e.g.* in a laminar flow hood) using correct aseptic techniques.
     - Handle antimicrobial disks with sterilized forceps and avoid touching surfaces.
7. **Problem: Delayed or inhibited fungal growth**
   - **Possible causes:**
     - **Suboptimal incubation conditions**: Incorrect temperature or humidity.
     - **Expired or poor-quality media**: Low availability of nutrients.
     - **Inhibiting components in the medium**: Presence of substances toxic to fungi.
   - **Solutions:**
     - Check and calibrate incubator settings to meet fungal growth requirements.
     - Use fresh, high-quality media prepared under sterile conditions.
     - Test for inhibitory components and switch to appropriate media formulations.

8. **Problem: Zones too large or disproportionate**
   - **Possible causes:**
     - **Agar too thin**: Accelerates diffusion of antimicrobial agent.
     - **Antimicrobial agent too strong**: High concentration or over-impregnation of disks.
   - **Solutions:**
     - Ensure agar thickness is uniform and meets standard guidelines.
     - Confirm the concentration of the antimicrobial agent and standardize the preparation of the disks.

References

1. Lee SC, Fung CP, Lee N, See LC, Huang JS, Tsai CJ, Chen KS, Shieh WB (2001) Fluconazole disk diffusion test with methylene blue- and glucose-enriched Mueller-Hinton agar for determining susceptibility of *Candida* species. J Clin Microbiol 39(4):1615–1617. https://doi.org/10.1128/JCM.39.4.1615-1617.2001

Spot Assay for Comparing Drug Susceptibility of Pathogenic Fungi in Agar

7

Abstract

A quantitative spot assay for fungi is a valuable technique that measures the number of colony forming units (CFUs) to accurately determine the concentration of viable fungal cells in a given sample. This method is particularly advantageous as it allows a qualitative assessment of fungal growth over a wide range of cell concentrations. This versatility makes it useful in various experimental contexts, including studying the effects of genetic mutations, drug treatments and different environmental conditions on fungal viability. The spot assay protocol is characterized by its speed, reliability and scalability, allowing researchers to efficiently study the dynamics of fungal growth in different experimental scenarios. This makes it an indispensable tool in the fields of fungal genetics and microbiology research as it provides insights into the complex interactions between fungi and their environment (Sahu et al, Bio Protoc 13:e4872, 2023). The topics covered in this protocol include several basic procedures, including preparing drug stocks, formulating yeast extract peptone dextrose (YEPD) agar plates (with and without drugs), culturing fungal strains, and performing serial dilutions. In addition, the analysis of the results is a crucial component that enables meaningful interpretation of the data. This includes the evaluation of growth patterns, the determination of the minimum inhibitory concentration (MIC) of drugs and the evaluation of the effects of specific genetic modifications on fungal growth. By integrating these methods, the quantitative spot assay provides a comprehensive framework for studying various aspects of fungal biology. Researchers can use this approach to gain a deeper understanding of fungal physiology, pathogenicity and resistance mechanisms, ultimately contributing to advances in both basic and applied mycological research. Overall, this technique serves as a cornerstone in the toolbox of mycologists and microbiologists and facilitates the study of fungi in various research applications (Letsiou et al, Appl Sci 14:1899, 2024).

R. Pasrija, D. Kumari, *Protocols for Studying Pathogenic Fungi*,
https://doi.org/10.1007/978-981-96-7047-5_7

Keywords

Quantitative · Spot assay · Fungal growth · YEPD · Drug

Introduction This chapter describes the protocols from culture preparation to spot assay performance. The following topics are covered as follows:

7.1. Materials Required
7.2. Preparation of Media and Buffers
7.3. Protocol
7.4. Tips and Troubleshooting

7.1 Materials Required

7.1.1 Fungal Strains and Growth Media

- **Fungal strains**: Pure cultures of the pathogenic fungi to be tested.
- **Growth media:** Agar medium suitable for fungi and drug susceptibility testing, such as Yeast extract peptone dextrose (YEPD), sabouraud dextrose agar (SDA) and potato dextrose agar (PDA).

7.1.2 Antifungal Drugs

- **Antifungal agents**: The drugs to be tested, such as azoles, echinocandins, or polyenes *etc*.
- **Stock solutions:** Prepare the drugs in appropriate concentrations in sterile solvent (*e.g.* dimethyl sulfoxide (DMSO) or sterile water).
- **Dilution materials:** Sterile water, saline, or broth to prepare serial dilutions of the drugs.

7.1.3 Preparation of the Fungal Suspension

- **Sterile saline solution or broth**: For the preparation of fungal suspensions (*e.g.* phosphate buffered-saline (PBS) or Roswell Park Memorial Institute (RPMI)-1640).
- **Spectrophotometer or hemocytometer:** For standardizing the fungal suspension to a specific cell density (*e.g.*, 1×10^6 cells/mL).

7.1.4 Agar Plates

- **Sterile Petri dishes**: Pre-poured with the selected growth medium.
- **Supplemented agar (if required):** For some assays, the agar may need to be supplemented with glucose, or other nutrients/drugs.

7.1.5 Spotting Tools

- **Pipettes and sterile tips**: Micropipettes for spotting precise amounts (*e.g.* 5–10 μL) of the fungal suspension onto the agar plate.
- **Template (optional):** A printed grid or template to guide the spotting positions.

7.1.6 Sterilization and Safety Materials

- **Bunsen burner or alcohol lamp**: For aseptic handling.
- **Autoclave:** For the sterilization of media and equipment.
- **Personal protective equipment (PPE)**: Gloves, lab coat, and safety goggles.

7.1.7 Incubation

- **Incubator**: Set to the optimum temperature for fungal growth (usually 30 °C).

7.1.8 Tools for Data Recording

- **Marker pens**: For labeling the plates with fungal strains, drug concentrations, and other details.
- **Notebook or spreadsheet software**: For recording observations, such as growth intensity or inhibition zones.

7.2 Preparation of Media and Buffers

7.2.1 Preparation of YEPD Media and Agar (See Chap. 2)

7.2.2 Preparation of PBS (See Chap. 3)

7.2.3 Preparation of Saline Solution (0.85%) (See Chap. 3)

7.2.4 Preparation of Drug Stocks (See Chap. 5)

7.3 Protocol

7.3.1 Inoculation of the Strains

- Streak all yeast strains of interest fresh from the glycerol stock onto a sterile YEPD agar plate and incubate at 30 °C for 48 h.
- Inoculate a single colony of each strain into 5 mL of autoclaved YEPD broth and incubate overnight at 30 °C with shaking at 200 rpm. Centrifuge at 3000–4000 rpm for 5 min, discard the supernatant and resuspend the cell pellet in 1 mL of sterile PBS to wash the cells.

7.3.2 Serial Dilution

- Dilute the washed culture in the spectrophotometer in 0.85% saline solution to OD_{600} 0.1.
- Label five sterile microcentrifuge tubes as 10^{-1}, 10^{-2}, 10^{-3}, 10^{-4}, and 10^{-5} (Fig. 7.1).
- Add 900 μL of sterile saline to each tube.
- Mix 100 μL of the 0.1 OD_{600} cell suspension with 900 μL of sterile saline in the first microcentrifuge tube. This results in a 1:10 dilution of the cells.
- Take 100 μL of the cell suspension from the first tube (10^{-1}) and add it to the second tube (10^{-2}). Mix thoroughly.
- Repeat the procedure to prepare the subsequent dilutions. For example, take 100 μL from the first tube (10^{-2}) and add it to the next tube (10^{-3}), and so on, until dilution 10^{-5} is made. Finally, discard the 100 μL cells from the fifth tube.

7.3.3 Spot the Dilutions onto Agar Plates

- Label the agar plate to indicate the dilutions to be spotted.

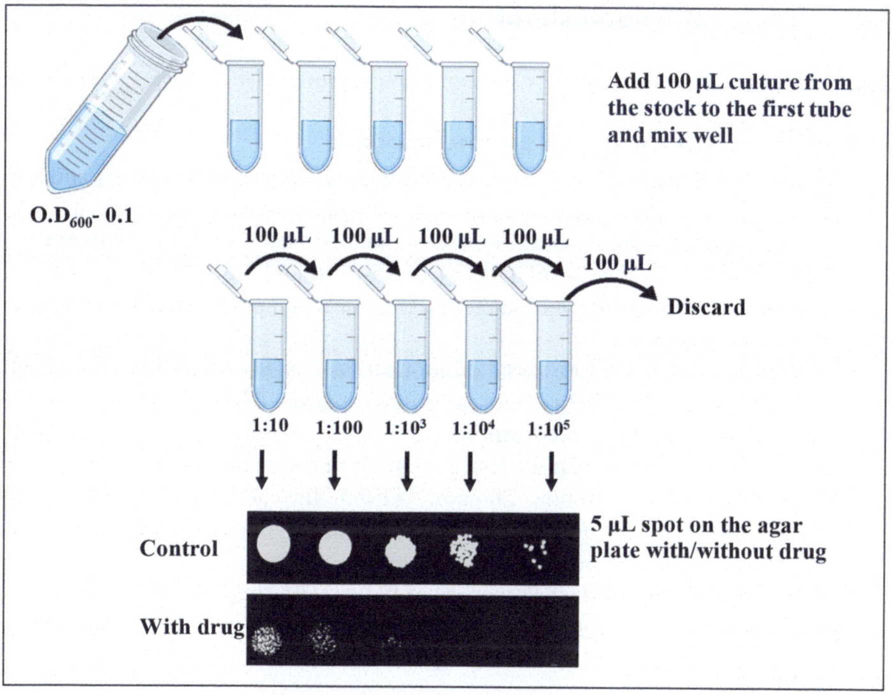

Fig. 7.1 Flow chart of the spot assay procedure for fungi

- Add 5 μL of each dilution to the agar plate, starting with the lowest dilution on the right side and moving to the highest dilution (towards the left side). This allows you to use a single tip, which minimizes the risk of carryover and prevents contamination of the higher dilutions with the cells of the lower dilution.
- Leave the plates undisturbed for 10–15 min after spotting to allow the culture to absorb fully. Then, incubate the plates at 30 °C for 48–72 h, depending on the growth observed (48 h for *Candida*/*Saccharomyces* sp. and 72 h for *Cryptococcus* sp).

7.3.4 Analysis

- Examine the plate after incubation for the growth of fungal colonies in each dilution.
- Compare the number of colonies at each dilution to assess viability or growth phenotype under different conditions.

7.4 Tips and Troubleshooting

7.4.1 Tips

1. **Fungal strain and preparation of the inoculum**
   - **Fungal strain selection:** Use freshly cultivated fungal strains that are actively growing and in the logarithmic phase. This ensures robust and uniform growth.
   - **Standardize the inoculum:** Prepare fungal suspensions with a constant concentration (*e.g.* 10^6 cells/mL) to ensure uniform and reproducible growth on the plate. Use a spectrophotometer (OD_{600}) or a hemocytometer to determine cell density accurately.
   - **Serial dilutions:** Prepare serial dilutions to assess growth at different fungal concentrations. This helps to assess dose-response effects.
2. **Preparation of the agar medium**
   - **Choose a suitable medium:** Use a medium that supports the growth of your fungal species. Common choices: YEPD for general fungal growth, RPMI-1640 agar for drug sensitivity testing and synthetic defined (SD) agar for nutrient-limiting conditions.
   - **Drug-containing plates:** Prepare agar plates with the desired drug concentrations. Use a control plate without the drug for comparison. Thoroughly mix the antifungal drug into the agar while it is still liquid to ensure even distribution.
3. **Spotting the fungal suspension**
   - **Spotting method:** Use a pipette or multichannel pipette to spot small, equal volumes (*e.g.* 5–10 μL) of each fungal dilution onto the agar surface. Make sure that the spots are equal and well-spaced so they do not overlap.
   - **Order of spotting:** Start with the least diluted sample to avoid carryover of higher cell concentrations.
4. **Incubation conditions**
   - **Temperature and time:** Incubate the plates at the optimum growth temperature for your fungus (*e.g.*, 30 °C for most fungi)
   - **Monitor growth:** Incubate for 48–72 h, depending on the fungal strain.
5. **Documentation and evaluation**
   - **Visual inspection:** Check for fungal growth in the spotted areas. Lack of growth indicates susceptibility to the drug at that concentration.
   - **Photographic records:** Take clear photographs of the plates for accurate comparison and documentation.
   - **Compare drug effects:** Compare growth on drug-containing plates with control plates to assess susceptibility.

7.4.2 Troubleshooting

1. **Problem: Inconsistent or absent growth of fungal colonies**

- **Possible causes:**
  - **Incorrect inoculation or dilution:** The cell suspension may have been diluted incorrectly or the inoculum may have been too small.
  - **Poor strain culture quality:** The fungal strain may not have been properly revived from the glycerol stock or may have been subject to genetic drift.
  - **Incorrect incubation conditions:** Incubation temperature or humidity may not be optimal.
  - **Expired or ineffective drug:** The drug stock may be degraded or ineffective due to improper storage or handling.
  - **Suboptimal medium composition:** The YEPD agar plate or broth may be contaminated or improperly prepared.
- **Solutions:**
  - **Check dilution accuracy:** Ensure that the cell suspension has the correct OD_{600} value (0.1) and that all dilutions are prepared carefully. Ensure that the correct colony count is achieved in the higher dilutions.
  - **Revive fungal strains properly:** Ensure that the yeast strains are freshly streaked from glycerol stocks and incubated at 30 °C for 48–72 h. To improve the viability of the strains, spread them several times if necessary.
  - **Provide the correct incubation conditions:** Check the temperature of the incubator (30 °C is optimal for yeast) and ensure that the plates are not overcrowded to provide sufficient space for the colonies to grow.
  - **Check the efficacy of the drug:** Ensure that the drug stock has been freshly prepared and confirm that it has been stored correctly at −20 °C. Use a control group to check that the drug affects growth as expected.
  - **Check the quality of the medium:** Prepare YEPD media and agar plates fresh or ensure they are not expired or contaminated. Consider re-autoclaving the media if contamination is suspected.

2. **Problem: Cross-contamination of drugs and dilution spots**
   - **Possible causes:**
     - **Wrong order of spots:** Spotting the fungal suspensions in the wrong order can lead to cross-contamination.
     - **Insufficient absorption time of the spots:** If you do not allow the spots to dry completely before incubation, the drug solutions may spread on the plate.
   - **Solutions:**
     - **Spot dilutions in the correct order:** Always spot the lowest dilution of the fungal culture first, and then gradually move to the higher dilution. This allows you to use a single tip, and ensure that the lowest dilution is transferred first, followed by the higher dilutions to avoid cross-contamination.
     - **Allow the entire absorption time to elapse:** Leave the plates undisturbed for 10–15 min after spotting to allow the culture to fully absorb before incubation.

3. **Problem: Inconsistent results between replicates**
   - **Possible causes:**
     - **Inconsistent preparation or handling of the drug:** If drug stocks are not prepared or stored properly, the results may vary between experiments.
     - **Variations in the fungal strain preparation:** If the yeast cultures are not standardized (*e.g.* inconsistent incubation times or different inoculum sizes), this can lead to discrepancies.
     - **Plate preparation issues:** Contaminated or improperly prepared plates can lead to inconsistencies between replicates.
   - **Solutions:**
     - **Standardize preparation of drug stocks:** Always prepare drug stocks at the same concentration and store them at −20 °C for consistent future use. Prepare fresh stocks if there are signs of deterioration.
     - **Ensure uniform preparation of inoculum:** Use the same procedure for inoculation, dilution and centrifugation each time.
     - **Ensure proper plate preparation:** Make sure all plates are prepared in a sterile environment and use the same media and reagents for each experiment.
4. **Problem: Difficulties in interpreting the results after a long incubation period**
   - **Possible causes:**
     - **Prolonged incubation distorts the results:** Prolonged incubation can cause yeast colonies to spread, making it difficult to distinguish between dilution spots.
     - **Insufficient growth for phenotypic assessment:** If growth is too slow due to drug exposure or other factors, it may be difficult to assess the growth phenotype.
   - **Solutions:**
     - **Optimize Incubation Time:** For longer-term assays, observe fungal growth at multiple time points (*e.g.* 24, 48 and 72 h). If you use longer incubation times, ensure that the plates are not overcrowded and that the colonies have room to form distinct clusters.
     - **Increase the sensitivity of the analysis:** In cases where colony numbers are difficult to assess visually, use a spectrophotometer or image analysis software to quantify colony growth or inhibition.

Time-Kill Kinetics to Determine the Number of Living Cells Over a Period of Time

8

Abstract

Time-kill kinetics in fungi is an important method for determining the rate at which fungal cells are eliminated over time in response to antifungal drugs or various stress conditions. This approach provides valuable insight into the dynamics of fungal cell death and the efficacy of antifungal treatments. The process usually begins by exposing fungal cultures to a specific antifungal drug or stressor, followed by sampling at predetermined time intervals, such as 0, 1, 4, 8, and 24 h (Öz et al, Mycopathologia 181:225–233, 2016). Techniques such as counting colony forming units (CFU) or performing viability assays, *e.g.* with methylene blue or propidium iodide, are used to quantify live cells. These methods allow researchers to accurately determine the number of viable cells remaining in the sample, which is crucial for calculating the rate of cell death over time. The data obtained from these experiments help to determine the time required to achieve a significant reduction in fungal cell populations and provide a better understanding of the pharmacodynamics of antifungal drugs. By analyzing time-kill kinetics, researchers can gain insight into the efficacy of antifungal treatments and thus make clinical decisions about dosing and treatment duration. In addition, this method sheds light on the underlying mechanisms of cell death, such as apoptosis or necrosis, and can reveal possible resistance mechanisms of fungal pathogens (Qureshi et al, In vitro and in silico approaches for the evaluation of antimicrobial activity, time-kill kinetics, and anti-biofilm potential of thymoquinone (2-Methyl-5-propan-2-ylcyclohexa-2,5-diene-1,4-dione) against selected human pathogens. 2022). The importance of time-kill kinetics goes beyond clinical applications. It is also important in agriculture and the environment, where fungal pathogens threaten hosts and ecosystems. By elucidating the effects of different antifungal strategies, this method contributes significantly to the broader understanding of fungal biology and

R. Pasrija, D. Kumari, *Protocols for Studying Pathogenic Fungi*,
https://doi.org/10.1007/978-981-96-7047-5_8

treatment strategies and ultimately helps in the development of more effective antifungal therapies (Adusei et al, J Trop Med 2019:1250645, 2019).

Keywords

Time-kill kinetics · Antifungal · CFU · Cell death · Efficacy

Introduction This chapter describes the protocols from culture preparation to the performance of time-kill kinetics. The following topics are covered as follows:

8.1. Materials Required
8.2. Preparation of Media and Buffers
8.3. Detailed Protocol for Time-Kill Kinetics
8.4. Tips and Troubleshooting

8.1 Materials Required

8.1.1 Fungal Culture and Medium

- **Fungal strains**: A pure culture of the fungus species to be tested.
- **Growth medium**: Liquid medium suitable for fungal growth (*e.g.* sabouraud dextrose broth (SDB), potato dextrose broth (PDB), or Roswell Park Memorial Institute (RPMI)-1640).
- **Agar plates**: Solid agar medium for colony forming unit (CFU) counts, yeast extract peptone dextrose (YEPD), sabouraud dextrose agar (SDA) and potato dextrose agar (PDA). Supplemented with antibiotics (*e.g.* chloramphenicol) to prevent bacterial contamination.

8.1.2 Antifungal Agents

- **Antifungal compound**: The drug or agent to be tested.
- **Stock solution**: Prepared in a suitable solvent (*e.g.* dimethyl sulfoxide (DMSO), ethanol, or sterile water) at a known concentration.
- **Diluent:** Sterile saline, water or broth to prepare working concentrations of the antifungal agent.

8.1.3 Materials for Sampling and Dilution

- **Sterile pipettes and tips**: For taking aliquots from the culture at specific intervals.
- **Sterile tubes**: For serial dilution of samples.

- **Dilution buffer**: Sterile saline solution or phosphate-buffered saline (PBS) solution for performing serial dilutions.
- **Vortex mixer:** For homogenizing the samples during the dilutions.

8.1.4 Incubation and Growth Monitoring

- **Incubator/shaker**: To keep the cultures under optimal growth conditions (*e.g.* temperature, stirring speed).
- **Timer or stopwatch:** To ensure precise sampling at set intervals.

8.1.5 Sterilization and Safety Materials

- **Bunsen burner or alcohol lamp**: For aseptic handling during sampling.
- **Autoclave**: For sterilization of media and equipment.
- **Personal protective equipment (PPE)**: Gloves, lab coat, and safety goggles.

8.1.6 Counting Tools/Devices

- **Petri dishes**: Pre-cast agar plates for plating diluted samples.
- **Colony counter (optional):** For easier and more accurate CFU counting.
- **Marker pen:** For labeling the plates with time points, dilution factor (DF) and other details.

8.1.7 Tools for Data Recording

- **Notebook or spreadsheet**: To record colony count and calculate CFU/mL over time.
- **Calculator**: To calculate CFU/mL based on DF and application volume.

8.2 Preparation of Media and Buffers

8.2.1 Preparation of YEPD Media (See Chap. 2)

8.2.2 Preparation of PBS and 0.85% Saline Solution (See Chap. 3)

8.2.3 Preparation of Yeast Nitrogen Base Medium (YNB) and Agar (See Chap. 4)

8.3 Protocol

8.3.1 Preparation of the Primary Culture

- Inoculate the fungal strain in YEPD and let it grow overnight at a suitable temperature (30 °C).
- Set the culture in RPMI-1640/YNB to $OD_{600} = 0.1$ to ensure that the same number of cells are inoculated into each tube.

8.3.2 Setting up the Plate

- Add 10 mL of your adjusted inoculum ($OD_{600} = 0.1$) to the 50 mL tubes.
- Add the antifungal agent (subminimal inhibitory concentration (MIC) or desired value) or stressor to the tubes, using untreated tubes as positive controls and tubes without inoculum as negative controls (see Chap. 5 for the protocol of the MIC test).

8.3.3 Incubation and Time Points

- Incubate the tubes at the desired temperature (*e.g.* 30 °C or 37 °C) under suitable conditions (*e.g.* shaking or static). **Note:** The incubation temperature for *Candida* and *Cryptococcus* sp. should be 30 °C. However, to study the effects of temperature fluctuations and the effects of shaking versus static conditions, cultures can be incubated at both 30 °C and 37 °C.
- Take samples from the tubes at specific time points (*e.g.* 0, 1, 2, 4, 6, 12, 24 h) to measure the number of live cells.

8.3.4 Dilution Series for Live Cell Counting

- Label sterile centrifuge tubes with the desired DF (*e.g.* 10^{-1}, 10^{-2}, 10^{-3}, *etc.*).At each time point, take 100 μL from the selected tubes and mix with the first tube containing 900 μL of saline (10^{-1} dilution). Shake briefly to mix.
- Transfer from tube 10^{-1} to the next tube (10^{-2}) containing 900 μL saline solution to obtain a tenfold dilution (as shown in Fig. 8.1). Shake briefly.
- Repeat this procedure for each additional 100 μL dilution to obtain a series of dilutions (up to 10^{-6}, depending on the initial concentration). Discard 100 μL from the last diluted tube.
- Add 5 μL of each dilution to an agar plate and incubate at 30 °C for colony formation (usually 48–72 h depending on the organism) (48 h for *Candida*/*Saccharomyces* and 72 h for *Cryptococcus* sp).

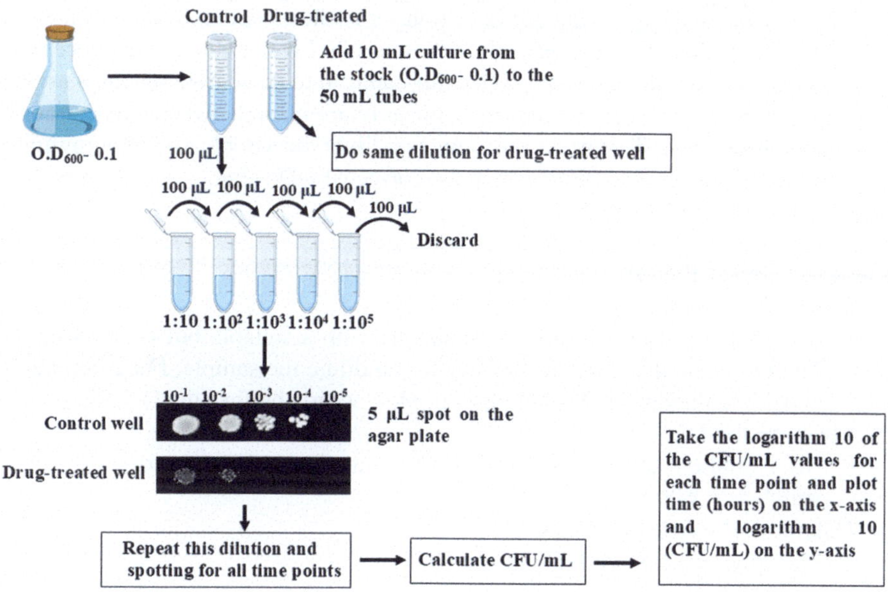

Fig. 8.1 Procedure of Time-kill kinetics

8.3.5 Counting the Colonies

- After incubation, count the colonies formed on the agar plates. Each colony represents a viable (living) cell. Use the following formula to calculate the CFU/mL:

$$\text{CFU / mL} = \text{Number of colonies} \times \text{dilution factor}\left(\text{DF}\right) / \text{Plated volume}\left(\text{in mL}\right)$$

- Take the 10 logarithmized values of CFU/mL for each time point.
- Plot the time (hours) on the x-axis and the logarithm 10 (CFU/mL) or CFU/mL on the y-axis to observe the time-kill kinetics.
- The resulting curve illustrates the time required to significantly reduce the number of living cells.
- Calculation example: If you obtain 2 colonies at a DF of 10^5:

  5 µL gave = 2 colonies, then 1000 µL cells would have = 2/5 × 1000 × DF (10^5) = 4 × 10^7 CFU/mL. The logarithm 10 of this CFU value would then be 7.602.

8.3.6 Statistical Evaluation

- Ensure that the experiment is performed with biological replicates (e.g. at least 3 independent experiments).
- Use software such as GraphPad Prism, R or Excel to analyze the data.

- Calculate the mean and standard deviation for each time point.
  - **NOTE:** If the OD_{600} in a growth curve exceeds 1 (often due to the concentration of the sample being too high), this can lead to inaccurate readings as OD values above 1.0 may not provide a linear relationship between concentration and absorbance. To correct this, you can dilute the sample and then multiply the OD value by the DF to obtain the corrected OD value.

You can find out how to do this here:

Steps to Correct the OD

1. **Dilute the sample**
   - If the OD exceeds 1.0, dilute the sample with a suitable buffer or medium. The DF value depends on how much you dilute the sample. For example, if you dilute the sample 1:10 (1 part sample + 9 parts buffer), the DF value is 10.
2. **Measure the OD**
   - After dilution, measure the OD of the diluted sample at the desired wavelength (*e.g.* 600 nm).
3. **Calculate the corrected OD**
   - Multiply the measured OD by the DF.

$$\text{Corrected OD} = \text{Measured OD} \times \text{DF}$$

- **Example**
  - Assume that the OD of your undiluted sample is 1.2 and is, therefore, outside the reliable range of your spectrophotometer.
  - You dilute your sample by a factor of 10 (*i.e.* 1:10 dilution).
  - You measure the OD of the diluted sample and find that it is 0.15.
  - Now, calculate the corrected OD:

$$\text{Corrected OD} = \text{Measured OD} \times \text{DF}$$

$$\text{Corrected OD} = 0.15 \times 10$$

$$\text{Corrected OD} = 1.5$$

The corrected OD of the undiluted sample is, therefore, 1.5.

8.4 Tips and Troubleshooting

8.4.1 Tips

1. **Preparation of fungal cultures**
   - **Use active cultures:** Start with a culture in the logarithmic phase to ensure that the cells are actively dividing, as these are the most sensitive to drugs. Inoculate a fresh culture and allow it to grow to the desired density (set $OD_{600} = 0.1$ for the first measurement).

- **Standardize the inoculum:** keep the initial cell concentrations constant for all replicates and experiments to ensure comparability.

2. **Selection and preparation of an antifungal drug**
   - **Concentration of active ingredient:** Use a range of concentrations to evaluate dose-dependent effects. Consider the following points:
     – A sub-MIC value for partial inhibition.
   - **Solubility of the drug:** Ensure that the drug is dissolved in a suitable solvent (*e.g.* DMSO, ethanol, or water) and diluted to the correct working concentrations.
3. **Time points/timing**
   - **Select appropriate time intervals**
     – Frequent sampling (*e.g.* every 30 min or 1 h) during the early phase of treatment.
     – Less frequent sampling (*e.g.* every 2–4 h) during the later phase.
   - **Define the endpoint:** Depending on the growth rate of the fungus, the test may take hours to days.
4. **Sampling and dilution**
   - **Uniform sampling:** Mix the culture thoroughly before sampling to ensure a representative cell distribution.
   - **Serial dilutions:** Perform serial dilutions of each sample to ensure countable colony count.
5. **Plating for viability**
   - **Choose the right medium:** Use nutrient-rich agar (*e.g.* YEPD, PDA, or SDA) that supports the growth of the target fungus.
   - **Plating technique:** Use a sterile spreader to spread the diluted sample evenly on the agar surface to avoid clumping, or use pipettes and tips to create round spots.
6. **Incubation**
   - **Temperature and duration:** Incubate the plates/tubes at the optimum temperature for the fungal species (*e.g.* 30 °C) for 48–72 h.
   - **Humidity control:** Ensure that the plates do not dry out during incubation by maintaining the correct humidity in the incubator.
7. **Colony count and data analysis**
   - **CFU count:** After incubation, count the colonies on the plates for each dilution. Multiply by the DF to determine the number of viable cells (CFU/mL).
   - **Plot the time-kill-kinetics curve:** Plot the number of viable cells (logarithm CFU/mL) against time to analyze the time-kill dynamics.

8.4.2 Troubleshooting

1. **Problem: Inconsistent colony counts or no colony growth**
   - **Possible causes:**

- **Incorrect preparation of the inoculum:** The inoculum may not have been set to the correct OD_{600} value, resulting in an inaccurate initial concentration of fungal cells.
- **Incorrect dilution procedure:** Serial dilution errors may result in an inaccurate dilution range affecting the colony count.
- **Contaminated agar plates:** If the agar plates or media are contaminated, fungal growth cannot be observed, leading to unreliable results.
- **Too high/low drug concentration:** If the concentration of the antifungal drug is too high, it can completely inhibit fungal growth and prevent colony formation. On the other hand, if the drug concentration is too low, there may be no measurable effect on the fungal population.

• **Solutions:**
  - **Ensure an accurate OD_{600} measurement:** Make sure the inoculum is set to $OD_{600} = 0.1$ to ensure a constant initial concentration. Use a spectrophotometer to confirm the OD prior to inoculation.
  - **Check the dilution technique:** Follow the dilution procedure carefully and ensure that each tube is mixed correctly and that accurate volumes are transferred. Re-check the DF (*e.g.* 10^{-1}, 10^{-2}, *etc.*) before plating.
  - **Check the quality of the plates and media:** Prepare fresh, sterile agar plates and ensure that the media is not expired or contaminated. Work under sterile conditions to avoid contamination.
  - **Optimize drug concentration:** Test a range of drug concentrations, including sub-MIC levels, to ensure that the concentration of antifungal drug is appropriate. Perform controls to confirm drug activity.

2. **Problem: Inaccurate or inconsistent timing of sampling**
   • **Possible causes:**
     - **Incorrect sampling procedure:** If samples are not collected correctly at a specific time point (*e.g.*, missed time points, inconsistent volumes), time-kill kinetics data will be unreliable.
     - **Insufficient mixing prior to sampling:** If the tubes are not mixed properly prior to sampling, the distribution of cells may not be uniform, resulting in inaccurate cell viability measurements.
     - **Incorrect incubation conditions:** Temperature or shaking conditions during incubation may not be optimal and affect fungal growth or drug activity.

   • **Solutions:**
     - **Ensure consistent sampling:** Collect samples at the specified time points (*e.g.*, 0, 1, 2, 4, 6, 12, 24 h) to track changes over time. Use a micropipette to ensure a consistent volume across all time points.
     - **Mix the tubes well before sampling:** Gently mix the tubes before sampling to ensure that the cells are evenly distributed.
     - **Optimize the incubation conditions:** Ensure that the incubation temperature is correct (*e.g.*, 30 °C) and that the plate is incubated under appropriate conditions (*e.g.*, shaking or static) to ensure uniform fungal growth and drug activity.

3. **Problem: The data does not show clear time-kill kinetics (no clear decrease in CFU)**
   - **Possible causes:**
     - **Insufficient time points:** If the time points are too far apart (*e.g.* critical early time points are missing), the change in CFU may not be recognized.
     - **Drug ineffectiveness:** The drug may not act on the fungal cells as expected, resulting in minimal changes in CFU.
     - **Uneven drug activity:** If the drug is not properly mixed or unevenly distributed the results may not show a clear pattern of time-kill kinetics.
   - **Solutions:**
     - **Increase the frequency of sampling:** Include more time points between key intervals (*e.g.* sampling at 0, 1, 2, 4, 6, 8, 12, and 24 h) to capture a more detailed time course of the fungal kill process.
     - **Test the efficacy of the drug:** Use controls to ensure that the drug is effective. If no significant effect is observed, either increase the drug concentration, check the stability of the drug stock, or ensure that the drug is active under the given conditions.
     - **Ensure uniform application of the drug:** Make sure that the drug is thoroughly mixed with the fungal cells so that each replicate receives the same drug concentration. If you are concerned about degradation of the drug, prepare fresh solutions before use.
4. **Problem: CFU counts are too high or too low**
   - **Possible causes:**
     - **Over-dilution or under-dilution:** Incorrect DF may result in too many or too few colonies on the plate.
     - **Incorrect plating volume:** If you plate too much or too little of the sample, the CFU calculations can be falsified.
   - **Solutions:**
     - **For high CFU numbers:** If CFU counts are too high, dilute the sample further (*e.g.* use dilutions of 10^{-6} or more).
     - **For low CFU counts:** If the CFU counts are too low, reduce the DF to increase the concentration.
     - **Ensure the correct volume for plating:** Always use the correct volume, usually 5 μL, to ensure consistency.
     - **Adjust the dilution range:** Dilute the sample to 10^{-10} to 10^{-15} and discard the previous dilution tubes. Only spot the last 5–6 dilution tubes to ensure that the colonies remain countable and are well distributed, especially at later time points. This helps to avoid overcrowding and ensures an accurate colony count.
5. **Problem: Problems with statistical analysis (variability between replicates)**
   - **Possible causes:**
     - **Insufficient biological replicates:** If there are too few replicates, the results can show considerable variability, leading to statistical errors.

  - **Inconsistent data collection or analysis:** Variability in data collection methods (*e.g.* inconsistent colony counts) or errors in the statistical analysis software could skew the results.
  - **Poor data quality:** If the CFU/mL values are inconsistent between replicates, the data may not be reliable for analysis.
- **Solutions:**
  - **Increase the number of biological replicates:** Ensure that the experiment includes at least three biological replicates to account for variability in fungal growth and response to the drug.
  - **Check the consistency of data collection:** Standardize colony counting process (count colonies at the same time after incubation, use the same method for counting).
  - **Use appropriate statistical software:** Use programs such as GraphPad Prism, R or Excel to calculate means and standard deviations. Ensure you use appropriate statistical tests (*e.g.* ANOVA, t-tests) to compare different conditions.

Checkerboard Analysis to Calculate the Fractional Inhibitory Concentration (FIC) Index of Pathogenic Fungi for Drug Interaction

9

Abstract

The emergence of antifungal resistance has led to the need for new strategies to combat fungal infections. One such strategy is the use of drug combinations that can potentially improve antifungal efficacy and circumvent resistance mechanisms. A common method for evaluating drug interactions is checkerboard analysis, which allows the determination of the FIC (Fractional Inhibitory Concentration) index, a quantitative measure of the interaction between two antifungal drugs (Bellio et al, MethodsX 8:101543 2021). In this method, different concentrations of two drugs are tested in a two-dimensional array of wells, with each combination tested over a range of concentrations. The FIC index (FICI) is calculated by evaluating the minimum inhibitory concentrations (MIC) of each drug in the combination, relative to its MIC when used alone (Meletiadis et al, Antimicrob Agents Chemother 54(2):602–609 2010). By testing different concentrations of two drugs in combination, we were able to identify the most effective combinations and evaluate their potential for synergistic effects. The FIC index shows considerable variability in the nature of the interactions, with some combinations showing strong synergistic effects, while others are additive, indifferent, or antagonistic. The results of the checkerboard analysis emphasize the potential of combination therapy in the treatment of fungal infections, particularly in pathogens that exhibit resistance to conventional antifungal agents. The studies, which focus on the molecular mechanisms underlying these interactions, could be crucial for the development of targeted combination therapies that can effectively counter the growing threat of antifungal resistance.

Keywords

Drug combination · Synergy · Antagonism · FIC · Lower concentration · Efficacy · Fungal infections · Fractional inhibitory concentration

R. Pasrija, D. Kumari, *Protocols for Studying Pathogenic Fungi*,
https://doi.org/10.1007/978-981-96-7047-5_9

Introduction This chapter provides a detailed protocol for checkerboard testing to analyze the FIC of antifungal drug combinations, from preparation to execution of the test. Tips and troubleshooting advice are also provided. The following topics are covered:

9.1. Materials Required
9.2. Preparations
9.3. Procedure
9.4. Tips and Troubleshooting

9.1 Materials Required

- **Microtiter plate**: A 96-well microtiter plate with a flat bottom
- **Fungal inoculum**: Spores or mycelium of the fungal pathogen
- **Growth medium:** RPMI-1640/Yeast Extract Peptone Dextrose (YEPD)/Potato Dextrose Broth (PDB)
- **Test compounds** (*e.g.*, antifungal agents/active ingredients)
- **Sterile distilled water or buffer**
- **Pipettes and tips**: Preferably multichannel pipette for more efficiency
- **Incubator**: For incubation the microtiter plate at a temperature suitable for fungal growth
- **Sterile tools** (*e.g.* microcentrifuge tubes, micropipettes, sterile glassware)
- **Plate reader**: For reading the absorbance

9.2 Preparations

9.2.1 Preparation of the Culture Media (See Chap. 2)

9.2.2 Preparation of the Fungal Culture (See Chap. 3)

9.2.3 Preparation of the Drug Stocks (See Chap. 5)

9.3 Procedure

Drug-drug interactions are investigated using a broth microdilution method based on checkerboard titration according to the recommendations of the National Committee for Clinical Laboratory Standards [1].

- This test uses the 96-well plate method and is similar in some respects to the MIC assay.

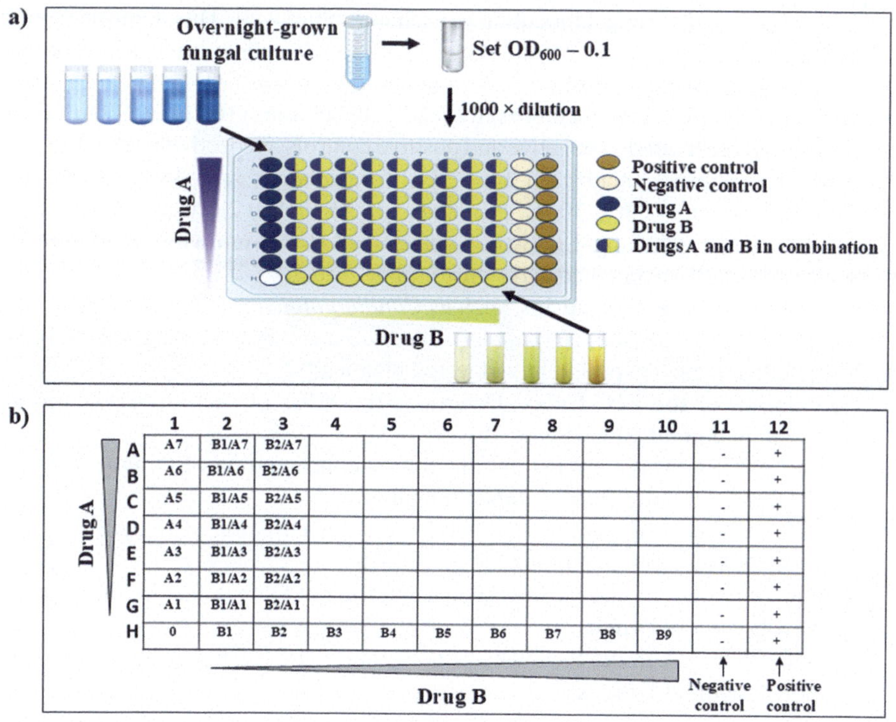

Fig. 9.1 (**a**) Illustration and (**b**) calculations of the checkerboard assay

- In the first step, the first column (from top to bottom) and the last row at the bottom (from right to left) of the microtiter plate are filled with 100 μL of two-fold serially diluted test drug A (in the column) and test drug B (in the row), respectively (Fig. 9.1).
- For the remaining wells, serially dilute Drug A and Drug B (starting with a four-fold higher concentration than the desired concentration) in separate reservoirs.
- Now, add 50 μL of the serially diluted drug A (from the reservoir) to wells along the ordinate (from A to G with the highest concentration at the top), while 50 μL of the serially diluted drug B (from the reservoir) is added to all wells along the abscissa (*i.e.* from tenth to second column with the highest concentration in the tenth column) (Fig. 9.1)
- Then adjust the concentration of the fungal culture in a spectrophotometer to 0.1 OD (0.1 OD corresponds to 1×10^6 cells/mL yeast cells); or 0.7 on the McFarland scale in the case of yeast (0.5 McF for molds) by adding sterile 0.9% saline solution. For yeast cultures, dilute the inoculum 1:1000 with sterile medium, for molds (such as *Fusarium* sp), dilute 1:50.

- Add 100 μL of this fungal inoculum to each microtiter well. This addition would help to achieve a final concentration of ~0.5–2.5 × 10^3 cells/mL in each well. For ~1 × 10^4 to 5 × 10^4 CFU/mL.
- The 11th (200 μL sterile medium only) and 12th columns (100 μL + 100 μL fungal culture) are treated as negative and positive controls, respectively (Fig 9.1a, b).
- After the required incubation time (*i.e.* for *Candida* sp.: 48 h at 30 °C, *Cryptococcus* sp.: 72 h at 30 °C, *Fusarium* sp.: 72–96 h at 28 °C and *Aspergillus* sp.: 48 h at 35 °C), fungal growth is determined at 595 nm using a microplate reader or visually using a concave mirror plate reader.
- Quality control (QC) organisms, such as *Candida albicans* SC5314, *Aspergillus fumigatus* ATCC 204305, *Cryptococcus neoformans* ATCC 208821, *etc.*, should be included in each experiment and tested identically.
- **Calculation of the FIC Index**: Determine the MICs of each drug when used alone and in combination (Fig. 9.1b):
  - MIC of Drug A in combination: the lowest concentration of Drug A in combination with Drug B that inhibits fungal growth.
  - MIC of Drug B in combination: the lowest concentration of Drug B in combination with Drug A inhibits fungal growth.
  - MIC of Drug A alone: the lowest concentration of Drug A that inhibits fungal growth without Drug B.
  - MIC of Drug B alone: the lowest concentration of Drug B that inhibits fungal growth without Drug A.

$$\textbf{FIC Index} = \mathrm{A / MIC\,A\left(FIC\,A\right) + B / MIC\,B\left(FIC\,B\right)}$$

- where A = concentration of drug A that is the lowest inhibitory concentration in its series, *i.e.* the MIC of A in combination, B = MIC of B in combination, and MIC A and MIC B are the MICs of drugs A and B alone, respectively.

- **Interpretation of the results:**
  - Synergistic interaction: FIC index ≤0.5 (indicating that the combination of drugs is more effective than either drug alone).
  - Additive interaction: FIC index >0.5 to 1.0 (indicating that the combination of the drugs is additive).
  - Indifferent interaction: FIC index 1 ≤ 4 (indicating that there is no significant interaction between the drugs).
  - Antagonistic interaction: FIC index ≥4 (indicating that the drugs attenuate each other's efficacy).
- **Analysis of the data:**
  - Plot dose-response curves for each drug combination.
  - Use statistical analysis to identify significant differences between drug interactions and controls.
  - Report the FIC index for each drug combination and classify the interaction accordingly.
- **Notes:**
  - Perform the test in triplicate to reduce variability and ensure reproducibility.

- Include positive and negative controls in each plate.
- Optimize and validate the assay conditions for each fungal species and drug combination.
- Use statistical analysis to confirm significant differences between drug interactions and controls.

9.4 Tips and Troubleshooting

9.4.1 Tips

1. **Prepare the drugs:**
   - **Select appropriate drugs:** Choose drugs with known antifungal activity and complementary mechanisms of action.
   - **Stock solutions:** Prepare drug stock solutions at concentrations that are multiples of the expected MICs.
     - Use sterile solvents such as DMSO or water, depending on solubility.
     - Ensure that the drugs are fresh and properly stored (*e.g.*, avoid repeated freeze-thaw cycles).
2. **Setting up the microtiter plates**
   - **96-well plates:** Use sterile, 96-well plates with a flat bottom to set up the checkerboard matrix.
     - Horizontal rows: Dilution rows of drug A.
     - Vertical columns: Dilution columns of drug B.
     - Combine both drugs in different concentrations in the wells, as shown in Fig. 9.1b.
   - **Preparation of the drug dilution: Double serial dilutions:** Prepare the serial dilutions of each drug independently in sterile medium. Ensure that the concentrations cover the range from below the MIC to above the MIC for both drugs.
   - **Final volume per well:** Maintain a uniform final volume (*e.g.* 100 μL per well).
   - **Preparation of the fungal inoculum: Standardize the inoculum:** Adjust the fungal suspension to a standard concentration using a spectrophotometer or hemocytometer.
   - **Even distribution:** Ensure that the inoculum is evenly mixed before dispensing to avoid variations.
3. **Incubation:**
   - **Optimal conditions:** Incubate the plates at the optimal temperature for fungal growth (*e.g.* 30 °C or 37 °C) for 24–48 h.
   - **Do not shake:** Do not shake the plates to avoid mixing the wells.
4. **Controls:**
   - **Wells for Positive control:** Add wells with each drug alone at different concentrations to determine individual MICs.

- **Wells for negative controls:** Include wells with media and fungal inoculum only to check for natural growth.
- **Empty control wells:** Media only (without inoculum) to account for background absorption.

9.4.2 Troubleshooting

1. **Problem: Irregular or inconsistent fungal growth**
   - **Possible causes:**
     - Inoculum concentration is too high or too low.
     - Expired or improperly stored fungal strains.
     - Suboptimal growth media or environmental conditions (*e.g.*, temperature, pH).
   - **Solutions:**
     - Prepare the fungal suspension at the recommended concentration (10^4–10^6 CFU/mL) and check with a hemocytometer or spectrophotometer.
     - Use freshly cultured fungal strains that have been grown on a suitable medium.
     - Ensure that the growth medium and incubation conditions (temperature, humidity, pH) meet the specific requirements of the fungal species.
     - Avoid exposing the plates to air for long periods of time to prevent drying out.
2. **Problem: Drug precipitation in the wells**
   - **Possible causes:**
     - Insufficient solubilization of the drugs.
     - Incompatibility of the solvent with the medium (*e.g.* DMSO or ethanol used in high concentrations).
   - **Solutions:**
     - Ensure that the drugs are completely dissolved in a suitable solvent (*e.g.*, DMSO for hydrophobic drugs).
     - Use $\leq 1\%$ DMSO or ethanol in the final well concentrations to minimize interference with fungal growth.
     - Sterile filter the drug solutions to remove undissolved particles.
3. **Problem: Cross-contamination between wells**
   - **Possible causes:**
     - Improper pipetting technique.
     - Splashing of liquid during handling or transportation of the plate.
   - **Solutions:**
     - Use a multichannel pipette with sterile, disposable tips for accurate dispensing.
     - Pipette carefully to avoid splashing and mix the contents thoroughly.
     - Close the plate with a lid or parafilm during incubation to prevent contamination.

4. **Problem: Edge effect on the microtiter plate**
   - **Possible causes:**
     - Evaporation of liquid from the edge wells.
     - Uneven temperature distribution in the incubator.
   - **Solutions:**
     - Use a humidified incubator to prevent evaporation.
     - Avoid using the outermost wells (leave them as blanks) or include additional controls to normalize.
     - Incubate plates in a stable, temperature-controlled environment.
5. **Problem: Inaccurate MIC determination**
   - **Possible causes:**
     - Subjective visual assessment of fungal growth.
     - Inconsistent distribution of inoculum.
   - **Solutions:**
     - Use an automated plate reader to measure OD instead of visual inspection.
     - Ensure even mixing of the fungal inoculum prior to distribution into the wells.
     - Repeat the test at least three times to account for variability.
6. **Problem: Unexpected drug interactions (antagonism instead of synergy)**
   - **Possible causes:**
     - Inappropriate drug combinations.
     - Decomposition of the drugs during incubation.
     - Interference between drugs and solvents or chemical interactions.
   - **Solutions:**
     - Check the stability of the drugs under experimental conditions (*e.g.* pH, temperature).
     - Test drugs individually to ensure they are active before proceeding with combinations.
     - Consider alternative combinations if antagonism is consistently observed.
7. **Problem: Poor reproducibility between experiments**
   - **Possible causes:**
     - Inconsistent preparation of fungal inoculum.
     - Variability of drug concentrations.
     - Changes in incubation conditions.
   - **Solutions:**
     - Standardize the preparation of fungal suspensions and ensure that CFU/mL remains constant between experiments.
     - Use freshly prepared, accurately diluted drug stocks.
     - Maintain identical incubation conditions (*e.g.* temperature, duration, and humidity) for all experiments.
8. **Problem: Excessive growth or contamination in the wells**
   - **Possible causes:**
     - Bacterial contamination or excessive growth of fungi.
     - Inappropriate aseptic techniques during plate preparation.

- **Solutions:**
  - Work under sterile conditions (*e.g.* in a laminar flow hood).
  - Use antibiotics in the growth medium if bacterial contamination is suspected.
  - Sterilize pipettes, tips and other equipment regularly

9. **Problem: High background OD or interference**
   - **Possible causes:**
     - Absorption of drugs or solvents at the measurement wavelength.
   - **Solutions:**
     - Use appropriate blank controls (wells with medium and drugs without fungal inoculum).
     - Subtract the background OD from the experimental readings before evaluating them.
10. **Problem: No growth is observed in the wells of the positive control.**
    - **Possible causes**
      - Incorrect preparation or dilution of the fungal inoculum.
      - Toxicity of the solvent used for the drugs.
    - **Solutions:**
      - Check the preparation of the fungal inoculum and ensure that the correct concentration is used.
      - Ensure that the concentration of the solvent (*e.g.* DMSO or ethanol) in the final wells does not exceed 1%.

References

1. National Committee for Clinical Laboratory Standards (1995) Reference method for broth dilution antifungal susceptibility testing for yeasts, Tentative standard. Document M27-T. NCCLS, Wayne, PA

Hyphal Induction in Yeasts and Molds

10

Abstract

Hyphal production is an important growth modification in yeasts and molds and illustrates the adaptive strategies of fungi to their environment. Yeasts, typically unicellular organisms, can develop hyphae or pseudohyphae in response to various environmental factors, including temperature fluctuations, nutrient deficiencies, or interactions with host organisms. A prime example is *Candida albicans*. It can produce both true hyphae and pseudohyphae mitotically without sexual reproduction in response to various environmental factors. Asexual reproduction (mitosis) is sufficient to produce these hyphae forms. Pseudohyphae cells are generally ellipsoidal (*i.e.* their width is greater in the middle than at the ends) and have constrictions at the septal junctions. In contrast, hyphal cells usually have parallel sides, are uniformly wide, and have septa without constrictions (Lu et al, Trends Microbiol 22(12):707–714 2007). On the other hand, *Cryptococcus neoformans* usually reproduces mitotically in the unicellular yeast form by budding. Under conditions that trigger sexual reproduction, such as dehydration/desiccation or nitrogen starvation, the fungus (cells of the opposite mating type α and a) undergoes the morphological transition from yeast to hyphae. Hyphal formation enables the uptake of nutrients from the environment and the production of stress-tolerant infectious spores (Chen et al, Virulence 11(1):337–348 2020). Since the transition from yeast to hyphae in *Cryptococcus* is clearly and closely linked to its sexual reproduction, it was not considered a classical dimorphic pathogen. This morphological change increases its virulence and its ability to invade host tissues, contributing to its role as an opportunistic pathogen. The ability of yeasts to switch between unicellular and filamentous forms emphasizes their remarkable plasticity and adaptability in different environments. In contrast, molds such as *Fusarium* sp. and *Aspergillus* sp. naturally exist in a multicellular, filamentous form consisting of hyphae. This structure is crucial for nutrient uptake, as it allows molds to efficiently absorb organic matter from their

R. Pasrija, D. Kumari, *Protocols for Studying Pathogenic Fungi*,
https://doi.org/10.1007/978-981-96-7047-5_10

environment and facilitate reproduction through sporulation. Unlike yeasts, hyphal growth in molds is not dependent on specific environmental triggers, but is a continuous and integral part of their life cycle. This distinction emphasizes the diverse growth patterns within the fungal kingdom, with hyphal formation fulfilling different functional roles depending on the organism. Understanding these growth dynamics is critical for both ecological studies and clinical applications, as the transition to hyphal growth in yeasts can significantly influence pathogenicity. Furthermore, this knowledge can inform strategies for the treatment of fungal infections and the development of antifungal therapies. Ultimately, the study of hyphal production not only reveals the complexity of fungal biology but also highlights the evolutionary adaptations that allow fungi to thrive in different niches (Chow et al, Pathogens 10:859, 2021; McGinnis, Tyring, Introduction to mycology. In: Baron S (ed) Medical microbiology. University of Texas Medical Branch at Galveston, Galveston, 1996).

Keywords

Candida · Hyphae · Virulence · Nutrient · Host interaction

Introduction This chapter describes protocols for hyphal induction in yeast. The topics covered include the basic protocol for hyphal induction in yeast.

10.1. Materials Required
10.2. Preparation of Media and Buffers
10.3. Protocol
10.4. Tips and Troubleshooting

10.1 Materials Required

10.1.1 Fungal Strains

- **Yeast cultures**: Pure culture of the species (*i.e. Candida albicans*) to be examined.

10.1.2 Growth and Induction Media

- **Liquid media:**
  - **Roswell Park Memorial Institute (RPMI)-1640**: with or without 3-(N-morpholino) propane sulfonic acid (MOPS buffer): Commonly used for the induction of *C. albicans* hyphae.
  - **Spider medium:** Specifically for the induction of hyphae in *C. albicans.*
  - **Yeast extract peptone dextrose (YEPD):** Can be used as a basal medium.

  - **Serum-containing media:** Add 10% fetal bovine serum (FBS) to RPMI-1640 or YEPD to induce yeast hyphae.
  - **Glucose-or nutrient-depleted media**: For starvation-induced hyphal growth.
- **Agar plates:** Solid spider agar/RPMI-1640 agar or YEPD agar for the formation of surface hyphae.

10.1.3 Materials for Environmental Control

- **Incubator**: Set it to specific temperatures: 37 °C for the induction of *C. albicans* hyphae. Optimal temperature for mold species (usually 25–30 °C).
- **CO_2 incubator (optional):** For yeast species that require special atmospheric conditions.

10.1.4 Sterile Materials for Sample Handling

- **Sterile water or saline solution**: For the preparation of fungal suspensions.
- **Pipettes and sterile tips**: For the precise handling of liquids.
- **Hemocytometer or spectrophotometer**: For standardizing the concentration of fungal cells/spores in 1 mL (*e.g.* 1×10^6 cells/mL).
- **Sterile tubes**: For preparation of the fungal inoculum and incubation.

10.1.5 Microscopy Accessories

- **Light microscope or phase contrast microscope**: For observing the formation of hyphae.
- **Microscope slides and coverslips/coverglasses**: For preparing fungal samples.
- **Staining agent (optional)**: Lactophenol cotton blue or calcofluor white for visualization of hyphae.

10.1.6 Safety and Aseptic Materials

- **Bunsen burner or alcohol lamp**: For aseptic handling.
- **Personal protective equipment**: Gloves, lab coat, and safety goggles.
- **Autoclave**: For the sterilization of media and equipment.

10.1.7 Tools for Time Measurement and Observation

- **Timer or stopwatch**: For monitoring incubation times.
- **Camera-compatible microscope or imaging system**: For documenting the development of hyphae.

10.2 Preparation of Media and Buffers

10.2.1 Preparation of YEPD +10% Serum

- First prepare YEPD media (see Chap. 2).
- **Add the serum:** Once the YEPD medium has cooled to approximately 50–55 °C, add the sterile FBS (or other suitable serum) to the medium. To obtain a final volume of 1 L, add 100 mL of serum to the cooled YEPD medium.
- **Mix thoroughly:** gently swirl or stir the mixture to ensure that the serum is evenly distributed in the medium.
- **Storage and quality control:** Store the liquid YEPD medium at room temperature and the agar plates upside down (to avoid condensation) at 4 °C.
- **Check for contamination:** Before using the YEPD media, incubate a sample of the liquid medium or a plate overnight at 30 °C to check for contamination. Signs of turbidity or unexpected growth indicate contamination.
- **Notes**
  1. Ensure that all materials and tools are sterile to avoid contamination.
  2. Label all containers with the date, contents, and additional instructions.
  3. The addition of serum improves the nutrient profile of the medium and can promote fungal growth, making this formulation useful for various experiments.

10.2.2 Preparation of RPMI-1640 + 10% Serum

- First prepare RPMI-1640 media (see Chap. 2).
- **Add serum:** Slowly add 100 mL of FBS (or the selected serum) to the RPMI-1640 medium while gently swirling or stirring to mix thoroughly.
- **Sterilize:** Filter the prepared medium through a 0.22 μm filter into a sterile bottle.
- **Aliquot and store:** Transfer the prepared RPMI 1640 + 10% serum medium into sterile bottles or flasks. Label the bottles with the date and contents. Store the medium at 4 °C for short-term use (up to 2 weeks). If you want to use it for longer consider aliquoting and freezing it at −20 °C or −80 °C.
- **Use:** Warm the RPMI-1640 + 10% Serum medium to room temperature or 37 °C before use, depending on the requirements of your experiment. Use this medium to induce hyphal growth in fungi by inoculating the desired fungal strain and incubating under appropriate conditions (*e.g.* temperature, aeration).
- **Notes**
  1. Always ensure that all materials are sterile to avoid contamination, especially when working with fungi.
  2. This protocol can be adapted according to the specific experimental requirements, *e.g.* the type of serum used or the volume prepared.

10.2.3 Preparation of PBS (See Chap. 3)

10.2.4 Preparation of Spider Medium

- Weigh the following quantities based on the desired final volume (*e.g.* for 1 L): 10 g mannitol (1%), 10 g Nutrient broth (1%), 2 g dibasic potassium phosphate (K_2HPO_4) (0.2%).
- Dissolve these components in distilled water to reach a final volume of 1 L.
- Stir the mixture thoroughly with a stirring rod or magnetic stirrer until all components are completely dissolved.
- **Check and correct the pH value (if necessary):** Although the spider medium normally works without the need for pH adjustment, however, pH can be checked with a pH meter or pH strip. The desired pH value is around 7.2.
- **Adjust the pH value (if necessary):** If the pH is outside the desired range, add a few drops of 1 N NaOH to raise the pH or HCl to lower the pH. Mix well after each adjustment.
- **Sterilize the medium:** Transfer the prepared spider medium to an autoclave-safe bottle or flask. Sterilize the medium by autoclaving it at 121 °C for 15 min. Allow the autoclave to cool down before carefully removing the bottle/flask.
- After cooling, store the spinning medium at 4 °C. It can be stored for several weeks.
- **Notes**
  1. Ensure all materials and tools are sterile to avoid contamination.
  2. Label all containers with the date and contents for easy identification.
  3. When ready to induce hyphal growth, inoculate the spider medium with the desired fungal strain and incubate under appropriate conditions (*e.g.* temperature, aeration).

10.3 Protocol

This protocol focuses exclusively on the methodology of hyphal induction in *Candida* sp. It does not address *Cryptococcus*, which requires sexual reproduction (mating between opposite a and alpha strains) to produce hyphae [1]. In addition, molds such as *Fusarium* sp. occur naturally in the environment in hyphal form. General culture preparations for spore suspensions from hyphal growth are discussed in Chap. 5.

10.3.1 Prepare the Yeast Culture

- Inoculate the yeast cells into the YEPD medium and incubate at 30 °C for 48 h to allow the yeast cells to grow.

- Collect the cells by centrifugation, wash them twice with PBS and suspend them at $O.D_{600}$ 0.4 in RPMI-1640/YEPD/Spider Media.

10.3.2 Inoculation in Hyphae-Inducing Media

- Inoculate the overnight grown yeast culture into a fresh hyphae-inducing medium, such as Spider medium, YEPD and RPMI-1640 supplemented with 10% serum

10.3.3 Incubation

- Incubate the cultures at 37 °C for 3–4 h. Ensure good aeration by keeping the culture volume low (5–10 mL in 50 mL flasks or tubes).
- Observe the cells microscopically at 1-h intervals to check for hyphae formation.

10.3.4 Microscopic Observation

- After 1–4 h, pipette 2–5 μL of the culture onto a clean glass slide.
- If necessary, dilute the culture with PBS or a suitable buffer to avoid overcrowding of cells and to ensure clear observation.
- Place a coverslip over the culture drop to smooth the sample and avoid air bubbles, that could obstruct the view.
- Set the microscope to a suitable magnification (usually 100× magnification is used for observation of yeast and hyphal morphology).
- Adjust the illumination and contrast so that the cells are clearly visible. Optimal brightfield or phase contrast microscopy settings should be used to better recognize the hyphal filaments.
- Focus on the sample at low magnification (*e.g.* 40×) and then switch to a higher magnification (100×) to view individual cells in more detail (as shown in Fig. 10.1).
- If necessary, use image analysis software (*e.g.* ImageJ) to quantify hyphal formation or measure hyphal length.
- Note the time at which hyphal formation begins and the percentage of cells that transition from yeast to hyphal form over time. You can observe the culture at 1-h intervals during the incubation period (*e.g.* 1, 2, 3 and 4 h).
- **Note**
  - If hyphal formation is not uniform or is delayed, check that the culture has been synchronized by ensuring all cells are in the same growth phase (logarithmic phase).
  - Different yeast species may require slightly different conditions for hyphal induction. The medium and environmental triggers can be optimized depending on the specific strain and research focus.

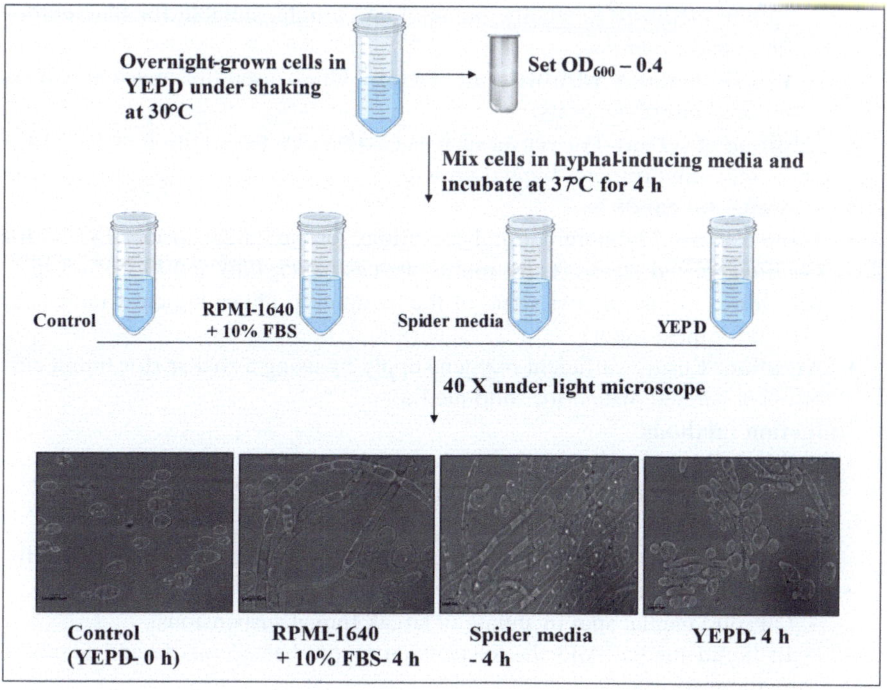

Fig. 10.1 Image showing hyphal induction and microscopy of *C. albicans* under different conditions at 37 °C. Cells in YEPD at 0 h served as a control, while the other media (RPMI-1640 + 10% FBS serum, spider medium and YEPD) were used as hyphal-inducing conditions

- Always include controls, such as yeast grown in YEPD at 30 °C, to compare yeast and hyphae forms.

10.4 Tips and Troubleshooting

10.4.1 Tips

1. **Selection of fungal strain and growth stage**
   - **Strain selection:** Use strains known for robust hyphae formation (*e.g. C. albicans*).
   - **Active cells:** Start with healthy, actively growing cells. Use cultures in mid-logarithmic phase as they respond better to induction signals.
   - **Standardize the inoculum:** Prepare a uniform cell suspension (*e.g.* 10^6 cells/mL) to ensure reproducibility.
2. **Selection of the induction medium**
   - **Defined media:** Use media that reliably induce hyphal formation, e.g.:
     - **RPMI-1640 (with or without serum):** Commonly used for *C. albicans*.

  - **Spider medium:** Contains mannitol and nitrate, suitable for filamentous fungi and *Candida*.
  - **YEPD medium with/without serum:** Often used for nutrient stress-induced morphogenesis.
- **Addition of serum:** For yeasts such as *C. albicans*, the addition of 10% FBS to the medium promotes hyphal growth.

3. **Environmental conditions**
   - **Temperature:** Maintain optimal conditions for the fungal species (37 °C for *Candida* sp. and 25–30 °C for molds such as *Aspergillus* and *Fusarium* sp).
   - **pH value:** Adjust the pH value of the medium to physiological values (*e.g.* pH 7.0 for most fungi).
   - **Aeration:** Ensure sufficient oxygen supply by using a shaker (for liquid cultures) or aerated plates (for solid media).
4. **Induction methods**
   - **Solid vs. liquid media:**
     - Try to test induction in both solid and liquid media.
     - Use agar plates to observe radial growth and hyphal branching.
     - Use liquid media to determine hyphal length or percentage of hyphal cells.
   - **Inoculation technique**
     - On solid media: spot inoculate or streak fungal suspensions.
     - In liquid media: Add the suspension to pre-heated media and incubate without stirring.
5. **Monitoring the development of hyphae**
   - **Time points:** Observe at regular intervals (*e.g.* 1, 2, 4, 6 h).
   - **Microscopy:** Light or fluorescence microscopy to confirm hyphae formation and to observe morphological characteristics.
   - **Quantification:** Count the hyphal cells compared to the yeast cells or measure the hyphal length using image analysis software.

10.4.2 Troubleshooting

Problem 10.1: No Hyphae Formation after Incubation

- **Possible causes:**
  - **Insufficient hyphae-inducing media**: The media used for induction may not contain the optimal conditions or nutrients for hyphae growth.
  - **Incorrect inoculum density**: If the yeast cells are not inoculated at the correct concentration (OD_{600}), they may not respond appropriately to the hyphae-inducing conditions.
  - **Insufficient incubation time or temperature**: The incubation time may be too short or the temperature too low to induce the hyphae formation.

- **Solutions:**
  - **Provide appropriate media**: Make sure that the hyphae-inducing media (*e.g.* YEPD with 10% FBS, Spider media, RPMI-1640 with serum) are fresh and properly prepared.
  - **Adjust the concentration of the inoculum**: Ensure that the yeast culture is suspended at an appropriate OD_{600} of 0.4 prior to inoculation.
  - **Increase the incubation time or temperature**: Ensure that the cultures are incubated at 37 °C (*Candida* sp) for 3–4 h and observe regularly under the microscope at 2–4-h intervals.

Problem 10.2: Hyphae Formation Is Delayed or Weak

- **Possible causes:**
  - **Incorrect aeration**: Insufficient aeration or insufficient culture volume can limit oxygen availability and hinder hyphae development.
  - **Overgrown or too-dense culture**: A very dense yeast culture can limit the availability of nutrients and prevent proper hyphal induction.
- **Solutions:**
  - **Provide adequate aeration**: Use low culture volumes (5–10 mL in 50 mL bottles or tubes) to maximize oxygen availability.
  - **Reduce cell density**: Use an appropriate inoculum concentration (OD_{600} of 0.4) to avoid overcrowding the culture, which can prevent proper hyphae formation.

Problem 10.3: Not all Cells Form Hyphae

- **Possible causes:**
  - **Uneven inoculum**: The initial inoculum on the plate may not be uniform, resulting in some cells not going through the induction process evenly.
  - **Insufficient growth time**: If the cells are not incubated long enough, they may not be synchronized or fully stimulated to form hyphae.
- **Solutions:**
  - **Ensure a uniform inoculum**: Start by making a uniform inoculum of cells. Use a sterile swab or pipette to spread or mix the cells evenly on the plate so that each cell has an equal chance to form hyphae.
  - **Synchronize the cells**: Incubate the cells for 48 h at 30 °C to ensure that the cells are fully synchronized. Make sure the culture medium is fresh and the environment is constant (temperature, humidity).

Problem 10.4: Difficulty Observing Hyphae Under the Microscope

- **Possible causes:**
  - **Improper sample preparation**: If the sample is not properly prepared (*e.g.* if the cells are too clumpy or diluted), it may be difficult to detect the formation of hyphae.
  - **Overexposure under the microscope**: Excessive exposure or incorrect magnification can make the hyphal filaments difficult to see.
- **Solutions:**
  - **Proper sample preparation**: Gently shake or properly resuspend the sample to avoid cell clumping. If the cells are too dense, dilute them slightly.
  - **Adjust the microscope settings**: Make sure the microscope is set to an appropriate magnification (usually 100×) to see the hyphal filaments. Use the correct light and contrast settings to improve visibility.

References

1. Liu Y, Wang R, Liu J, Fan M, Ye Z, Hao Y, Xie F, Wang T, Jiang Y, Liu N, Cui X, Lv Q, Yan L (2024) The vacuolar fusion regulated by HOPS complex promotes hyphal initiation and penetration in *Candida albicans*. Nat Commun 15(1):4131. https://doi.org/10.1038/s41467-024-48525-5

Isolation of Genomic DNA from Pathogenic Fungi

11

Abstract

Extraction of pure genomic DNA from fungi such as *Candida* sp., *Saccharomyces* sp./*Cryptococcus* sp. and molds like *Fusarium* sp. is essential for a wide range of molecular biology research applications, such as polymerase chain reaction (PCR), genotyping and sequencing. These different species present varied challenges in DNA extraction, particularly due to structural differences in their cell walls. In particular, *Cryptococcus* sp. possesses a thick polysaccharide capsule that must be effectively removed during the extraction process, requiring a more rigorous approach to ensure complete lysis. In contrast, other fungi such as *Candida* sp./*Saccharomyces* sp. do not have capsules, so their extraction is relatively simple and primarily requires mechanical disruption and cell lysis buffer for efficient cell wall degradation (Tapia et al, Mol Biotechnol 33(1):67–70 2010). However, molds such as *Fusarium* sp., pose an additional challenge due to their rigid and multilayered chitinous cell wall, which is often thicker and more resistant than yeasts. These complex cell wall components require a more aggressive lysis approach to achieve complete cell wall degradation. Our protocols utilize mechanical disruption by bead beating, which applies sufficient force to break down the robust cell wall structure, followed by the application of lysis buffers designed to degrade both cell wall components and intracellular proteins. This technique is founded on the phase separation principle, in which phenol-chloroform is used to separate lipids and proteins in the organic phase and isolate nucleic acids in the aqueous phase. To obtain high-quality genomic DNA, the extraction protocol includes precipitation steps with ethanol or isopropanol, effectively concentrating the DNA, while removing impurities. This optimized approach ensures efficient recovery of high-quality genomic DNA and significantly increases the reliability of downstream analyses (Codreanu, Ciurea, Microorganisms 11(4):818, 2023, Bolano et al, FEMS Yeast Res 1(3):221–224, 2001).

R. Pasrija, D. Kumari, *Protocols for Studying Pathogenic Fungi*,
https://doi.org/10.1007/978-981-96-7047-5_11

Keywords

Genomic DNA · Extraction · *Candida* · Capsule · Cell wall · Ethanol · Phenol-chloroform

Introduction This chapter describes the protocols from reagent preparation to the detailed DNA extraction steps in pathogenic fungi. The following topics are covered as follows:

11.1. Materials Required
11.2. Preparations
11.3. DNA Extraction Procedure for *Candida/Saccharomyces* sp.
11.4. Procedure for DNA Extraction for *Cryptococcus* sp.
11.5. Procedure for DNA Extraction for *Fusarium/Aspergillus* sp.
11.6. Gel Electrophoresis
11.7. Tips and Troubleshooting

11.1 Materials Required

11.1.1 Fungal Strains and Culture Media

- **Fungal strains**: Pure cultures of pathogenic fungi from which genomic DNA needs to be isolated.
- **Growth media:**
  - Liquid or solid growth medium for culturing the fungi, such as:
    - Sabouraud dextrose broth (SDB)
    - Yeast extract peptone dextrose (YEPD)
  - Agar plates (if required): For plating and isolating colonies of fungal cells.

11.1.2 Materials for the Collection of Fungal Cells

- **Sterile pipettes and tips**: For transferring fungal cultures.
- **Centrifuge tubes:** For obtaining fungal cells from the culture medium.
- **Centrifuge:** For pelleting the fungal cells after cultivation.

11.1.3 Reagents for Disrupting the Cell Wall

- To efficiently disrupt the tough fungal cell wall, we need the reagents that physically or chemically disrupt the cells.

- **Mechanical disruption agents:** Glass beads for beating beads in a vortex mixer is required for tougher fungi.
- **Detergents:**
  - **Sodium Dodecyl Sulfate (SDS):** To disrupt the fungal cell membrane.
  - **Cetyltrimethylammonium Bromide (CTAB):** Commonly used in protocols for DNA extraction from fungi.

11.1.4 Reagents for DNA Extraction

- **Extraction Buffer**: Usually consists of Tris-HCl (for buffering), ethylenediaminetetraacetic acid (EDTA) (for chelating divalent ions and protecting DNA), NaCl (for ionic strength), SDS (for cell membrane lysis) or CTAB (depending on the protocol).
- **Phenol-chloroform-isoamyl alcohol:** Used for phase separation during DNA extraction.

11.1.5 DNA Precipitation and Purification

- **Ethanol/isopropanol**: For genomic DNA precipitation after extraction.
- **70% ethanol**: For washing the DNA pellet to remove residual salts.
- **TE buffer (10 mM Tris-HCl, 1 mM EDTA, pH-8)/MQ**: For resuspending the DNA after precipitation.
- **Note**: TE buffer is generally used for prolonged storage of DNA, as it maintains DNA stability over longer periods. The Tris component helps maintain a stable pH. At the same time, EDTA is a chelator that forms complexes with divalent cations (such as Mg^{2+}) essential cofactors for nucleases, thus preventing the degradation of DNA or RNA by nucleases. However, for certain downstream applications, such as PCR, it is recommended to store the DNA in 10 mM Tris-HCl (without EDTA). The reason for this is that EDTA chelates Mg^{2+} ions, which are essential for the activity of polymerase in PCR. The lack of sufficient magnesium ions can severely impair the function of the polymerase and reduce the efficiency and success of the PCR amplification.

11.1.6 Materials for Quality Control

- **Nanodrop spectrophotometer or microplate reader:** To check A_{260}/A_{280} ratio.
- **Agarose gel electrophoresis setup**: Agarose powder (typically 0.8–1.2% gel), gel loading dye (*e.g.*, ethidium bromide (EtBr)) for DNA visualization and electrophoresis chamber and power supply for gel electrophoresis.

11.1.7 Safety and Aseptic Materials

- **Bunsen burner or alcohol lamp**: For aseptic handling, especially in cultures.
- **Sterile pipette tips and tubes**: To transfer reagents without contamination.
- **Personal protective equipment (PPE)**: Laboratory coats, protective eyewear and gloves.
- **Autoclave or sterilizer**: For sterilizing media, tools, and equipment.

11.2 Preparations

11.2.1 Preparation of YEPD Media (See Chap. 2)

11.2.2 Preparation of PDB (See Chap. 2)

11.2.3 Preparation of the Phenol-Chloroform-Isoamyl Alcohol Solution (PCI)

- **Ingredients:** Isoamyl alcohol, 1 M Tris-HCl, chloroform, Tris-saturated phenol, vortex mixer, storage container (preferably amber-colored glass), *etc.*
- **Preparation of Tris-saturated phenol**
  - If pre-saturated phenol is not available, start by saturating phenol with Tris:
  - For this, mix phenol with an equal volume of 1 M Tris-HCl (pH 8.0).
  - Allow the solution to separate into two layers.
  - Use only the upper (aqueous) layer, containing the phenol saturated with Tris-HCl.
- **Note:** Always work with phenol in a properly ventilated environment, as it is toxic and has a strong odour.
- **Preparation of PCI solution**
  - Mix phenol, chloroform and isoamyl alcohol in a clean container in a ratio of 25:24:1.
  - Mix the components by briefly vortexing to ensure the solution is homogeneous.
- **Storage:** Store the PCI solution in an amber-colored glass container at 4 °C.
- **Tips for using the PCI solution**
  - **Avoid exposure to light**: The PCI solution is sensitive to light. Therefore, always keep it in an opaque container or cover it with aluminum foil to prevent degradation.
  - **Phase separation after use**: After mixing, allow the PCI solution to separate into two layers: the upper aqueous layer and the lower organic layer. Make sure that you use the lower layer for your extraction

11.2.4 Preparation of the TAE Buffer

- **Ingredients:** EDTA, Tris-base, glacial acetic acid, distilled water, beaker, pH meter, autoclave, *etc*.
- **Preparation procedure**
  - To prepare 100 mL of 50-fold TAE solution, following ingredients need to be mixed: Tris-base (24.2 g), glacial acetic acid (5.71 mL) and 0.5 M EDTA solution of pH 8.0 (10 mL) and finally adjust the final volume to 100 mL. Autoclave it for sterilization.
  - Store the TAE buffer in an amber-colored bottle to protect it from light, and store it at 4 °C/room temperature.

11.2.5 Preparation of Breaking Buffer

- **Prepare the components**
  - **Triton X-100 (2%)**: To dissolve cell membranes. Pipette the required volume for a final concentration of 2%.
  - **SDS (10%)**: This anionic detergent denatures proteins and degrades cellular components.
  - **5 M NaCl:** Sodium chloride is used for the ionic strength, necessary to disrupt cell membranes and break down protein-protein interactions.
  - **1 M Tris-HCl (pH 8.0):** Tris-HCl is a buffering agent that helps to keep the pH stable during the lysis process.
  - **0.5 M EDTA (pH 8):** EDTA functions as a chelator, interacting with divalent cations (such as Mg^{2+} and Ca^{2+}) and thus prevents their participation in enzymatic activities (*e.g.* DNases, RNases).
- **Mixing the breaking buffer**
  - In a clean flask, add 2 mL of Triton X-100 (for 2% working concentration), 10 mL of 10% SDS (working concentration—1%), 2 mL of 5 M stock NaCl (working concentration—100 mM), 1 mL of 1 M stock Tris-HCl (pH 8) (10 mM working concentration) and 2 µL of 0.5 M EDTA (pH 8) (1 mM working concentration).
  - After adding all components, dilute the solution with water to make 100 mL.
- **Sterilize by filtering**
  - Sterilize the buffer through a 0.22 µm filter to remove all potential impurities and contaminants.
  - This ensures the solution is sterile and suitable for biological applications.
- **Storage**
  - Store the prepared breaking buffer at room temperature in a sealed container, preferably in a dark bottle to avoid degradation of the components.
  - Make sure to label the container with the preparation date and contents.

11.2.6 Preparation of Urea Buffer

- **Ingredients:** Urea, NaCl, Tris-HCl, SDS, EDTA, deionized water, NaOH or HCl, autoclave
- **Buffer components**
  - **Urea (chaotropic agent)**: Urea is a chaotropic agent that helps to denature proteins and destroy cellular structures. It also facilitates the denaturation of proteins and stabilizes nucleic acids during extraction.
  - **NaCl (salting-out agent)**: Sodium chloride maintains the ionic strength of the solution, which contributes to the stability of DNA and facilitates the separation of proteins and nucleic acids.
  - **Tris-HCl (buffering agent, pH 8.0)**: Tris-base serves as a pH stabilizer to keep the pH at 8.0, which is optimal for DNA extraction and helps stabilize DNA.
  - **SDS (SDS detergent)**: SDS is a detergent that dissolves the membrane of cells and denatures proteins.
- **0.5 M EDTA (pH 8.0, chelating agent)**: EDTA prevents nucleases from degrading DNA by binding divalent cations, that are crucial cofactors for nuclease activity.
- **Procedure for the preparation**
  - **Dissolve the components**
    - In a clean beaker or container, add about 700 mL distilled water, 480 g urea (8 M working concentration), 29.2 g NaCl (0.5 M working concentration), 2.42 g Tris-base (20 mM working concentration) and 20 g SDS (2% working concentration).
    - Stir the mixture thoroughly so that it dissolves completely. Due to its solubility, it may take some time for the urea to dissolve and heat up during dissolution.
    - After the solids are dissolved, add 40 mL of 0.5 M EDTA (20 mM working concentration) (pH 8.0). Stir the mixture well so the EDTA is completely mixed with the solution.
    - Adjust the pH to 8.0 by adding 1 N NaOH or HCl drop by drop with constant stirring until the pH reaches 8.0.
    - After the pH is adjusted, adjust the solution volume to 1 L using distilled water. Mix thoroughly to obtain a homogeneous solution and autoclave. Autoclaving ensures the buffer is free of microbial contamination and is safe for subsequent biological applications.
  - **Storage**
    - Let the buffer cool at room temperature after autoclaving.
    - Store it in a sealed beaker (room temperature) and shield it from light to avoid degradation of the components.

11.2.7 Preparation of Lysis Buffer

- **Components of the lysis buffer**
  - **SDS (0.5%)**: SDS disrupts cell membranes by dissolving lipids and denaturing proteins. It is commonly used in DNA extraction protocols for effective cell lysis.
  - **Sarkosyl (0.5%)**: Sarkosyl is a stronger detergent that aids in the breakdown of cell walls and membranes and increases the efficiency of cell lysis.
  - **TE buffer (pH 7.5)**: The TE buffer contains 10 mM Tris for buffering and 1 mM EDTA to chelate divalent metal ions (which protect DNA from degradation by nucleases).
- **Preparation procedure**
  - **Dissolve the components**
    - In a clean beaker or container, dissolve 5 g SDS (to achieve a final concentration of 0.5%) and 5 g Sarkosyl (to achieve a final concentration of 0.5%) in 500 mL of TE buffer.
    - Stir the mixture thoroughly so that the solids dissolve completely. This may take some time, especially for SDS, as it takes longer to dissolve in cold buffers.
    - Use 1 N NaOH to bring the pH to 7.5 or HCl, drop by drop, with constant stirring. This is important to ensure the buffer has the correct pH for optimal DNA lysis and protection.
    - After the pH is adjusted to 7.5, make the solution up to 1 L by adding distilled water. Stir thoroughly to obtain a homogeneous solution.
  - **Sterilization**
    - Autoclave it 121 °C for 15 min. Autoclaving ensures the lysis buffer is free of microbial contamination, making it safe for downstream biological experiments.

11.2.8 Preparation of CTAB Extraction Buffer (2%)

- **Ingredients**
  - CTAB (2 g), NaCl, EDTA (stock solution—0.5 M; pH 8.0), Tris-HCl (stock solution—1 M; pH 8.0), distilled water.
  - Optional:
    - β-Mercaptoethanol: 1% (add just before use; it reduces disulfide bonds in proteins and helps to neutralize phenolic compounds that could interfere with nucleic acid extraction.).
    - Polyvinylpyrrolidone (PVP): 0.2% (binds to and removes polyphenolic compounds, ensuring cleaner and less contaminated nucleic acid extraction).

- **Preparation**
  - **Prepare stock solutions**
    - 1 M Tris-HCl solution: Dissolve 121 g of Tris-base (121 g) in 1 L of distilled water.
    - EDTA (0.5 M stock solution): Dissolve disodium EDTA (186.1 g) in 1 L of distilled water.
  - **Mix the buffer components**: 2 g CTAB (2 g/L working concentration), 81.76 g NaCl (1.4 M working concentration), 10 mL 1 M Tris-HCl stock (100 mM working concentration) and 4 mL0.5 M EDTA stock (20 mM working concentration). Add distilled water to bring the volume to ~900 mL.
  - **Dissolve the solution and adjust the volume**
    - Heat the solution gently (at ~60 °C) with stirring to dissolve CTAB completely.
    - Top up the solution with distilled water to reach 1 L and adjust to pH 8.0.
  - **Storage**
    - Store at room temperature or 4 °C for long-term use.
    - Warm the buffer to 65 °C before use to ensure that CTAB is completely dissolved.
  - **Notes**
    - Fresh additives: If using β-mercaptoethanol or PVP, add them just before use.
    - Ensure proper labeling of the buffer and safety measures, especially when handling β-mercaptoethanol.
    - Store the CTAB buffer protected from light and contamination.

11.2.9 Preparation of Phenol-Chloroform (1:1)

- Add equal quantities of phenol and chloroform to a suitable container and mix.
- Check and correct the pH value to 8.0.
- **Note:** Always handle phenol and chloroform in a fume hood, and wear suitable personal protective equipment.

11.2.10 Preparation of RNase A Stock Solution

- Determine the amount of RNase A powder required.
- Weigh 1 mg of RNase A powder using an analytical balance.
- Shift the powder to a microcentrifuge tube.
- Add 1 mL DNase-free MQ water or a suitable buffer (*e.g.* TE buffer).
- Blend the solution by repeatedly pipetting up and down or by gently inverting the tube.
- Label the tube with the concentration (1 mg/mL), date, and other relevant information.
- Store the solution at −20 °C for prolonged use or at 4 °C for short-term keeping.

11.2.11 Preparation of 0.8% Agarose Gel

- **Ingredients:** Agarose powder, 1× TAE buffer (If you have a 50× TAE stock, dilute to 1× with distilled water), microwave or hot plate, comb and gel casting tray to prepare the gel, pipette and tips to measure, EtBr, weighing balance to measure agarose powder.
- **Preparation of the agarose solution**
  - To prepare a 0.8% agarose gel, calculate the amount required for the desired gel volume, *e.g.* for 100 mL, you need 0.8 g agarose is needed.
  - Add the agarose powder in a container (*e.g.*, glass beaker).
  - Add 50 mL of the 1× TAE buffer (or the required volume) to the beaker with the agarose powder (0.4 g).
- **Heat the solution**
  - Heat the agarose-TAE solution in the microwave or hotplate to completely dissolve the agarose. If using a microwave, heat the solution in short bursts (20–30 s), swirling gently between each burst to prevent boiling over.
  - Once the solution is clear (no floating containing agarose particles), allow it to cool slightly. The temperature should be comfortable to the touch but not hot. At this point, mix 3–6 μL of the 10 mg/mL EtBr solution.
- **Pour the gel**
  - Pour the mixture in the casting tray with caution and fill to the desired level.
  - Allow the gel to cool until it solidifies (at room temperature).

11.3 DNA Extraction Procedure for *Candida/Saccharomyces* Sp.

- **Cell growth:** Inoculate 5 mL of YEPD broth with *Candida* or *Saccharomyces* cells and place overnight at 30 °C overnight in shaking incubator.
- **Cell harvest:** Centrifuge the culture at 5000×*g* for 5 min to pellet the cells. Discard the supernatant carefully.
- **Cell resuspension:** Gently resuspend the pellet in 200 μL of breaking buffer using a pipette.
- **Cell lysis:** Add an equal volume (200 μL) of glass beads (0.5 mm) and 200 μL PCI (lower phase) to the tube. Shake vigorously for 5 min to disrupt the cell walls. Ensure that the sample is thoroughly mixed.
- **Phase separation:** Add 200 μL of 1× TE buffer and spin the mixture at 5000×*g* for 5 min at room temperature to facilitate phase separation.
- **Transfer the aqueous phase:** Without disturbing the interphase, upper DNA containing phase (aqueous) has to transfer gently into a microcentrifuge tube.
- **RNA removal:** Add 3–5 μL RNase A (1 mg/mL) in the DNA containing vial and keep the tube at 37 °C for 30 min to degrade the RNA present.
- **Second phase separation:** Add an equal amount of PCI to the tube. Shake the mixture thoroughly. Spin at 5000×*g* for 5 min at room temperature. Repeat this step if necessary to ensure a clean separation.

- **Final transfer of the aqueous phase:** Carefully collect the upper phase again into a new tube.
- **DNA precipitation:** Add almost double the quantity of cold isopropanol or 100% ethanol to the aqueous phase and invert gently to mix. Let the solution sit at room temperature for 5 min to precipitate the DNA.
- **Pellet the DNA:** Store the mixture at −20 °C for at least 30 min to enhance DNA precipitation. Centrifuge at 10,000×*g* for 15–20 min at 4 °C to pellet the DNA.
- **Wash the DNA pellet:** Carefully remove the supernatant and wash the pellet with 500 µL of 70% ethanol. Centrifuge at 10,000×*g* for 5 min, then discard the ethanol. Allow the DNA pellet to air dry at room temperature for 5 to 10 min.
- **DNA resuspension:** Resuspend the DNA pellet in 50 µL DNase-free MQ water or TE buffer. Store the genomic DNA at −20 °C for long-term preservation (as shown in Fig. 11.1). Refer to sect. 11.1.5 for guidelines on storing DNA in TE buffer, either with or without EDTA [1, 2].

11.4 Procedure for DNA Extraction from *Cryptococcus* Sp. (Fig. 11.2)

- **Cell growth:** Inoculate 15 mL of YEPD medium with *Cryptococcus* cells. Incubate the culture under shaking at 150 rpm (overnight) at 30 °C. Ensure that the culture is well-aerated for optimal growth.
- **Harvest cells:** Post-incubation, spin the culture at 4000×*g* for 5 min. Discard the supernatant without damaging the cell pellet. Wash the cell pellet twice with cold

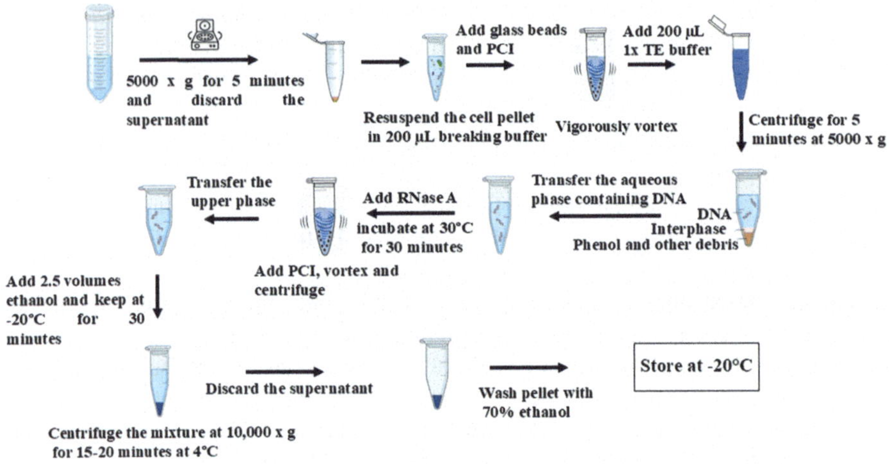

Fig. 11.1 Figure shows the DNA extraction protocol for *Candida*/*Saccharomyces* sp.

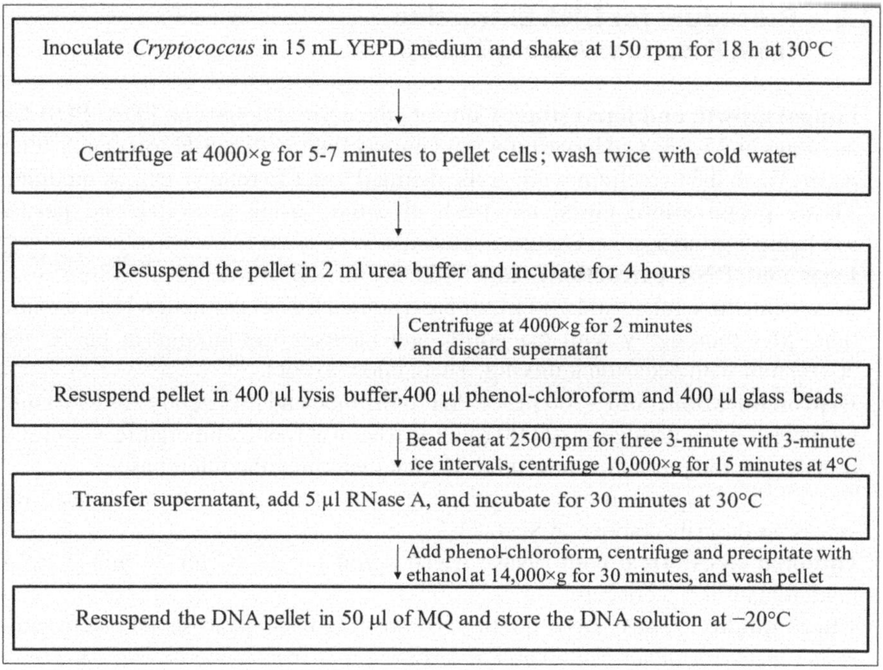

Fig. 11.2 Flowchart of the DNA extraction protocol from *Cryptococcus* cells

water to remove residual media and nutrients. Centrifuge at 4000×*g* for 5 min between washes.

- **Treatment with urea buffer:** Resuspend the washed pellet in 2 mL of urea buffer and incubate at 30 °C or room temperature with gentle agitation for 4 h. Following incubation, centrifuge the solution at 4000×*g* for 2 min and carefully discard the supernatant, ensuring the pellet remains undisturbed.
- **Cell lysis:** Dissolve the pellet in 400 µL of lysis buffer. Add 400 µL phenol-chloroform (lower phase) to the tube, followed by 400 µL acid-washed glass beads. Mix the contents thoroughly.
- **Bead beating:** Perform bead beating by running the vortex mixer at 2500 rpm for three cycles of 3 min each, with 3-min breaks on ice at 4 °C between cycles. Centrifuge the lysate after bead beating at 10,000×*g* for 15 min at 4 °C. Carefully transfer the supernatant to a new microcentrifuge tube without disturbing the pellet.
- **RNA removal:** Mix 5 µL of RNase A (1 mg/mL) into the supernatant and keep at 37 °C for 30 min to degrade the RNA in the sample.
- **Phase separation:** Combine with an equal volume of phenol-chloroform (1:1) to the supernatant. Spin at 2500 rpm for 3 min. Gently transfer the upper layer in a new tube.
- The remaining steps are identical to those outlined in Sect. 11.3.

11.5 Procedure for DNA Extraction from *Fusarium*/*Aspergillus* Sp.

- **Fungal growth and harvesting:** Cultivate the desired fungus on PDA/ PDB for 5–7 days at 25–28 °C. Harvest the fungal mycelium with sterile forceps or filter paper. Wash the mycelium with sterile distilled water to remove excess medium.
- **Tissue preparation:** Grind the fresh mycelium using a mortar and pestle/ mechanical grinder.
- **Lysis and DNA extraction:** Add ~100 mg of the powdered mycelium to a microcentrifuge tube. Add 700 μL of pre-warmed CTAB extraction buffer to the tube. Mix thoroughly with the vortex and incubate the mixture at 65 °C for 30–60 min with occasional mixing. Then, allow to cool.
- **Deproteinization:** Add ~700 μL of PCI to the tube and mix gently for 5–10 min by inversion. Centrifuge at 12,000 rpm for 10 min at room temperature. Carefully transfer the top phase to a new tube without disturbing the interphase.
- The DNA precipitation and resuspension steps are identical to those used for yeasts, as described above in Sect. 11.3.
- **Optional steps:** Treat with RNase A (10 μg/mL) at 37 °C for 30 min if RNA contamination is a concern.
- **Check quality:** Assess DNA quality and quantity using agarose gel electrophoresis (check for intact bands) and absorbance in a spectrophotometer (A_{260}/A_{280} ratio ~1.8 indicates good purity) [3].

11.6 Gel Electrophoresis

- **Ingredients:** DNA sample, bromophenol blue dye (tracking dye), 1× TAE buffer (for running the gel), agarose gel (prepared according to the desired concentration), electrophoresis apparatus (gel box, power supply), micropipettes, gloves and tips, *etc.*
- **Prepare the bromophenol blue dye**
  - **Mix 30% glycerol with a pinch (1–2 mg) of bromophenol blue dye**
    - **30% glycerol**: Increases the density of the sample so that it can settle in the wells of the gel.
    - **Bromophenol blue**: Serves as a tracking dye to visually monitor the migration of the DNA.
    - Mix thoroughly until the dye is completely dissolved.
- **Prepare the DNA sample**
  - Mix the DNA sample with the bromophenol blue loading dye.
  - The final volume should be 15 μL.
- **Load the gel**
  - The already prepared agarose gel must now be loaded in the electrophoresis chamber.
  - Add the prepared 15 μL of DNA sample with the loading dye to each well.
  - Take care not to overload the wells.

- **Set up the electrophoresis device**
  - **Set the gel:** Ensure that the gel is immersed in 1× TAE buffer.
  - **Connect the electrodes**
    - **Red cable (positive)**: Connect it to the anode (positive side) of the electrophoresis chamber.
    - **Black cable (negative)**: Connect it to the cathode (negative side) of the electrophoresis chamber.
    - This orientation ensures that the negatively charged DNA migrates toward the positive electrode (anode) (as shown in Fig. 11.3).
- **Run the gel**
  - Set the power supply for the electrophoresis to 60 V to start the run.
  - At 60 V, the DNA travels slowly, ideal for larger fragments.
- **Monitor the progress**
  - As the current flows through the electrophoresis buffer, bubbles begin to form. The formation of these bubbles indicates that the current is flowing correctly, and the electrophoresis is proceeding as expected.
  - Check the migration of the tracking dye (bromophenol blue) as it moves through the gel. You can use this dye to track how far the DNA has travelled.

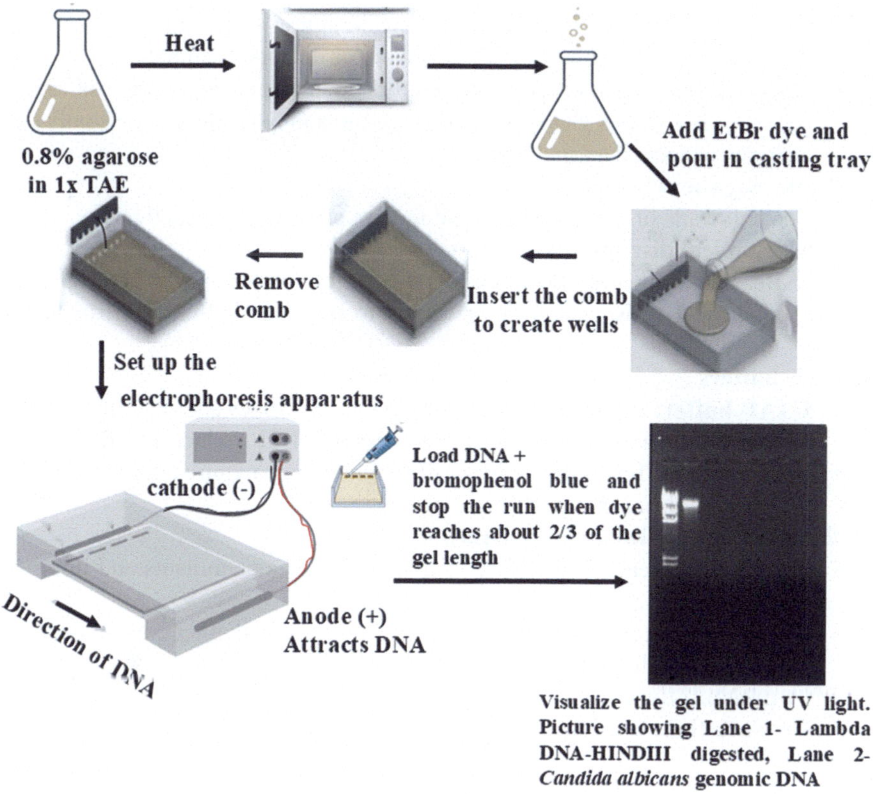

Fig. 11.3 Preparation of the agarose gel and the agarose gel electrophoresis procedure

- **Stop the run and visualize**
  - When the bromophenol blue dye has reached about 2/3rd of the gel length, stop the electrophoresis so the dye does not run out of the gel.
  - Carefully take the gel out of the electrophoresis chamber.
  - Visualize the gel under UV light to check for DNA bands.
- **Keynotes**
  - The red cable (positive) is connected to the anode (positive side), while the black wire (negative) is connected to the cathode (negative side).
  - The bubbles that appear in the buffer indicate the current flow. This is a sign that the electrophoresis is working properly.
  - The bromophenol blue dye moves in front of the DNA and indicates the progress of the electrophoresis. Monitor the gel carefully to ensure the DNA does not run out or over-migrate.

11.7 Tips and Troubleshooting

11.7.1 Tips

1. **Starting material**
   - **Fresh or frozen culture:** Use fresh fungal cultures or frozen samples stored that have been stored at −80 °C. Avoid old or degraded samples.
   - **Optimal cell density:** Grow the culture to the mid-logarithmic phase to achieve a high DNA yield. Overgrown cultures may produce excess polysaccharides or secondary metabolites.
2. **Rupturing the cell wall**
   The cell walls of fungi are rigid and require effective lysis methods:
   - **Mechanical disruption:**
     - **Bead beating:** Use glass beads in a bead mill or vortex with lysis buffer.
     - **Chemical lysis:** add SDS or CTAB to the lysis buffer to disrupt lipid membranes.
3. **Lysis buffers**
   - **CTAB buffer:** Effective for removing polysaccharides and other impurities. Use in combination with β-mercaptoethanol or DTT to reduce disulfide bonds.
   - **SDS-based buffer:** Suitable for most fungi but may require additional purification for polysaccharide-rich species.
4. **Removal of proteins and impurities**
   - **RNase treatment:** Add RNase to remove RNA contaminants and improve DNA purity.
   - **Organic extraction:** Use PCI to separate proteins and other contaminants.
   - Ensure thorough mixing and centrifugation to achieve clear phase separation.
5. **DNA precipitation**
   - **Precipitation:** Use isopropanol or ethanol to precipitate the DNA.

6. **Evaluation of quality and quantity**
   - **Purity check:** Measure the DNA purity with a spectrophotometer (the A_{260}/A_{280} ratio should be 1.8–2.0).
   - **Integrity check:** Run a small aliquot on an agarose gel to confirm high molecular weight DNA.

11.7.2 Troubleshooting

1. **Problem: Incomplete lysis of fungal cells**
   - **Possible causes:**
     - **Cell wall composition:** Fungal cell walls are complex and composed of chitin, glucans and mannans, making them resistant to lysis, especially in *Cryptococcus* sp. with an additional polysaccharide capsule.
     - **Capsule interference:** The *Cryptococcus* capsule (consisting of polysaccharides) provides additional protection to the cell, making it difficult for the lysis buffer or detergents (SDS, Sarkosyl) to effectively break down the cell wall and release the DNA.
   - **Solutions:**
     - **For *Candida* sp. (not capsulated):** Increase the concentration of detergents (e.g. SDS) in the lysis buffer if the initial lysis is incomplete. Extend the incubation time with lysis buffer to ensure thorough lysis.
     - **For *Cryptococcus* (encapsulated):** Use a more concentrated lysis buffer, possibly adding 0.5% sarkosyl or a higher concentration of SDS. Increase the bead-beating cycle to 6–8 times.
2. **Problem: Contamination of the DNA with RNA or protein**
   - **Possible causes:**
     - **Incomplete removal of RNA:** During the extraction process, especially within fungal cells, the RNA can co-extract with the DNA.
   - **Solutions:**
     - Use an RNase treatment during the DNA extraction protocol to degrade the RNA and minimize RNA contamination. Increase the RNase concentration and incubation time. Remember that RNase works efficiently at 37 °C.
     - Ensure the lysis buffer contains protease inhibitors to prevent protein degradation during cell lysis.
     - Perform an additional phenol-chloroform extraction step to further purify the DNA solution.
3. **Problem: Low DNA yield**
   - **Possible causes:**
     - **Inefficient cell lysis:** If the fungal cells are not completely lysed, the DNA yield is low. This problem can be more pronounced in species such as *Cryptococcus*, where the capsule hinders complete lysis.
     - **DNA degradation:** Prolonged contact with nucleases or improper storage conditions can lead to DNA degradation.

- **Solutions:**
  - Ensure complete lysis by optimizing the concentration and incubation time of the lysis buffer.
  - Reduce the size of the cell pellet by harvesting fewer cells or reducing the size of the starting culture. This will ensure that the volume of lysis buffer is sufficient to fully lyse all cells in the pellet.
  - Use bead beating or mechanical disruption methods to help break up the cell wall.
  - Avoid prolonged storage of the lysate or work quickly to avoid DNA degradation. EDTA in the lysis buffer chelates divalent cations and inhibits nucleases.
  - After phenol-chloroform extraction, store the upper aqueous phase (containing the DNA) at −20 °C overnight to improve DNA precipitation. The cold temperature promotes aggregation of the DNA and enables more efficient precipitation. Centrifuge the sample the next day at high speed (*e.g.*, 12,000–14,000 rpm) for 20–30 min to pellet the DNA.

4. **Problem: Poor DNA quality (fragmentation)**
   - **Possible causes:**
     - **Degradation due to prolonged lysis or handling:** Prolonged incubation or improper handling can lead to DNA fragmentation.
     - **Heat denaturation**: Excessive heating during extraction can denature the DNA, resulting in fragmented material.
   - **Solutions:**
     - Keep the lysis process and incubation times short and at controlled temperatures
     - Minimize handling and mechanical stress after DNA extraction.
     - Avoid repeated freeze-thaw cycles to prevent DNA degradation.
5. **Problem: Difficulties with DNA quantification**
   - **Possible causes:**
     - **Presence of impurities:** Polysaccharides (especially from *Cryptococcus* capsules) or proteins can interfere with accurate DNA quantification, especially when using spectrophotometric methods (*e.g.* UV-Vis at 260/280 nm).
   - **Solutions:**
     - Use RNase treatment and/or phenol-chloroform purification to remove impurities before quantification.
     - Perform gel electrophoresis to verify the integrity of the DNA and the absence of contaminants before performing spectrophotometric quantification.
     - Good quality DNA usually has an A_{260}/A_{280} ratio of about 1.8, but a ratio below 1.8 indicates protein and phenol contamination in the sample. To improve DNA purity, perform an additional phenol-chloroform extraction to remove proteins or perform an ethanol wash to remove residual impurities.

– If the ratio is above 2.0, this indicates DNA denaturation or RNA contamination. In this case, perform an RNase treatment to degrade the RNA, followed by a phenol-chloroform extraction or an additional ethanol wash to remove residual RNA and other contaminants to improve the quality of the DNA sample for subsequent applications.

References

1. Dymond JS (2013) Preparation of genomic DNA from *Saccharomyces cerevisiae*. In: Lorsch J (ed) Methods in enzymology. Academic Press, London, pp 153–160
2. Lõoke M, Kristjuhan K, Kristjuhan A (2011) Extraction of genomic DNA from yeasts for PCR-based applications. BioTechniques 50(5):325–328. https://doi.org/10.2144/000113672
3. Umesha S, Manukumar HM, Raghava S (2016) A rapid method for isolation of genomic DNA from food-borne fungal pathogens. 3 Biotech 6(2):123. https://doi.org/10.1007/s13205-016-0436-4

Exopolysaccharide Secretion by Pathogenic Fungi

12

Abstract

The secretion of exopolysaccharides (EPS) by pathogenic fungi plays a crucial role in their virulence and survival within host organisms. These complex carbohydrate structures are secreted into the extracellular environment and contribute significantly to the ability of fungi to establish and maintain infections. One of the main functions of EPS is to facilitate biofilm formation, a process that allows fungal cells to adhere to surfaces and form protective communities. Biofilms increase the resistance of fungi to physical and chemical stress factors and enable them to withstand host immune responses. In addition, EPS acts as a barrier that prevents the penetration of antifungal drugs and thus increases the resistance of fungi to treatment. This resistance is particularly problematic in clinical settings where biofilm-associated infections are difficult to eradicate. The structural components of EPS also play a role in modulating the host's immune response and help fungi evade detection and destruction (Malinovská et al, J Fungi 9:955, 2023). By masking pathogen-associated molecular patterns, EPS can reduce the effectiveness of immune responses and allow the fungi to survive and proliferate in the host. In addition to these protective functions, EPS contributes to nutrient acquisition by facilitating interactions with the environment and thus supporting the fungus' general metabolic processes (Snarr et al, J Fungi (Basel) 3:47, 2017). The synthesis of EPS is regulated by various environmental factors, including nutrient availability and stress conditions, indicating its dynamic role in fungal physiology. Understanding the mechanisms behind EPS secretion and its impact on pathogenicity is essential for developing new therapeutic strategies. By targeting EPS production or function, it may be possible to increase the efficacy of antifungal treatments and improve clinical outcomes for patients with fungal infections (Mahapatra and Banerjee, Microbiol Insights 6:1–16 2013). Overall, EPS is an important factor in the complex interplay between pathogenic fungi and their hosts, influencing both disease progression and response to treatment

R. Pasrija, D. Kumari, *Protocols for Studying Pathogenic Fungi*,
https://doi.org/10.1007/978-981-96-7047-5_12

(Wang et al, Heliyon 10(12):e32766, 2024). This method describes the procedure for the extraction of cell-bound and cell-free EPS from fungal strains and the quantification of EPS by the phenol-sulfuric acid method (Abirami et al, Front Microbiol 11. https://doi.org/10.3389/fmicb.2020.561298 2020; Nielsen, Total carbohydrate by phenol-sulfuric acid method. In: Nielsen food analysis laboratory manual. Springer, Cham, 2017).

Keywords

EPS · Host · Biofilm · Stress · Cell-bound · Cell-free

Introduction This chapter describes protocols for the extraction, storage and quantification of exopolysaccharides (EPS). The following topics are covered as follows:

12.1. Materials Required
12.2. Preparation of Media and Buffers
12.3. Protocol for EPS Extraction and Storage
12.4. Quantification of EPS After Storage
12.5. Tips and Troubleshooting

12.1 Materials Required

12.1.1 Fungal Strains and Growth Media

- **Fungal strains:** Pure cultures of pathogenic fungi (*e.g. Candida* sp./*Cryptococcus* sp).
- **Growth media:** Liquid media
  - **Yeast Extract Peptone Dextrose (YEPD)**: For general fungal growth.
  - **Sabouraud Dextrose Broth (SDB)**: Commonly used to cultivate fungi.
  - **Minimal Medium (MM)** (optional): For controlled EPS production, often supplemented with simple sugars such as glucose, galactose, or sucrose.
- **Agar plates (if required):** YEPD agar/SDB agar plates (for colony growth and isolation of fungal strains).

12.1.2 Materials to Stimulate EPS Secretion

- **Carbon sources**
  - Glucose (common carbon source for fungal growth).
  - Sucrose, fructose, maltose, galactose (alternative sugars to study their effects on EPS production).

- **Nitrogen sources**
  - Ammonium sulfate or ammonium nitrate.
  - Peptone or yeast extract.
  - To test for nitrogen deficiency or nutrient limitation.
- **Osmotic stress inducers (if required)**
  - Sodium chloride (NaCl), potassium chloride (KCl), or mannitol to induce stress that can promote EPS production.

12.1.3 Materials for Isolation of EPS

- **Centrifuge tubes:** 15 mL or 50 mL sterile tubes for harvesting fungal cultures or supernatants.
- **Centrifuge:** This is used to pellet the fungal cells after culture (usually at 3000–5000×*g* for 5–7 min).
- **Filtration materials:** Sterile filters (*e.g.*, 0.22 μm) are used to remove cells and debris from the culture supernatant (optional).
- **Ethanol or isopropanol:** For precipitation of exopolysaccharides from the culture supernatant.

12.1.4 Materials for the Characterization of EPS

- **Phenol-sulfuric acid reagent:** To quantify the total carbohydrate content of EPS (via the phenol-sulfuric acid assay).
- **Gas chromatography-mass spectrometry (GC-MS) equipment (optional):** To analyze the monosaccharide composition of EPS.
- **Fourier transform infrared spectroscopy (FTIR) (optional):** To determine functional groups and structural features of the EPS.

12.1.5 Safety and Aseptic Materials

- **Sterile pipettes and tips:** For accurate transfer of fungal cultures and reagents.
- **Sterile tubes and flasks:** For following aseptic techniques when handling fungal cultures.
- **Bunsen burner or alcohol lamp:** For maintaining aseptic techniques during fungal culture handling.
- **Personal Protective Equipment (PPE):** Gloves, lab coat, and goggles for safety during handling and experimental procedures.
- **Autoclave or sterilizer:** For sterilizing media, tools, and equipment.

12.2 Preparation of Media and Buffers

12.2.1 Preparation of YEPD Media (See Chap. 2)

12.2.2 Preparation of PBS (See Chap. 3)

12.2.3 Preparation of Isotonic Buffer

Materials Required Tris solution, NaCl solution, ethylenediaminetetraacetic acid (EDTA) solution, distilled water, HCl, NaOH, pH meter, autoclave, *etc.* for the preparation of 10 mM Tris, 0.9% NaCl, 1 mM EDTA.

- **Procedure:**
  - Weigh 1.21 g Tris base to obtain 10 mM Tris, 9 g NaCl to obtain 0.9% NaCl and 0.372 g EDTA to obtain 10 mM EDTA in 1 L.
  - Pour about 900 mL of distilled or deionized water into a beaker. Slowly add the 1.21 g Tris base to the water, stirring constantly, until it completely dissolves. Then add the 9 g NaCl and continue stirring until completely dissolved. Finally, add 0.372 g EDTA and stir until completely dissolved.
  - Use a pH meter to measure the pH of the solution. Adjust the pH to 7.4 with HCl or NaOH as needed. Add small amounts, stir well and check the pH again until the desired value is reached.
  - Transfer the solution to a volumetric flask and add distilled water to bring the total volume to 1 L.
  - Autoclave the buffer at 121 °C for 15 min or filter it into sterile bottles using a 0.22 μm filter.
  - Store the buffer at 4 °C for short-term use or at −20 °C for long-term storage. Label the container with the composition of the buffer, the date of preparation and any additional instructions.
  - **Notes:** Ensure all glassware are clean to avoid contamination. The buffer can be adjusted to different volumes as required while maintaining the same molar ratios.

12.3 Protocol for EPS Extraction and Storage

12.3.1 Materials Required

- Fungal strain, YEPD, PBS, centrifuge tubes, centrifuge, incubator, Isotonic buffer and ice-cold ethanol (100%), *etc.*

12.3.2 Procedure

- **Preparation of the fungal culture and cell harvesting**
  - Inoculate the pathogenic fungal strain in YEPD broth and incubate the cultures at 37 °C with shaking for 16–18 h.
  - Centrifuge the collected cells at 8000×*g* for 10 min.
  - Store the cell-free culture supernatant (CFCS) from each sample at 4 °C.
- **Preparation of the cell-bound EPS**
  - Resuspend the cell pellets in freshly prepared isotonic buffer/sterile PBS.
  - Use a bead mill or bead beater to physically break up the cells. This can be done by placing the fungal cells in a bead tube with glass beads and vortexing or shaking for 10–15 min.
  - Incubate the suspensions overnight at 4 °C to release the cell-bound EPS.
- **Collection and precipitation of the cell-bound EPS**
  - After overnight incubation, centrifuge the cell suspensions at 12,000 rpm for 15 min to collect the supernatant containing the cell-bound EPS.
  - Combine the supernatant with the cell-bound EPS with the previously stored CFCS of the respective samples.
  - Add 3 volumes of ice-cold ethanol to each combined mixture of cell-free and cell-bound EPS.
  - Incubate the mixtures overnight at −20 °C to allow precipitation of the EPS.
- **EPS collection, storage and analysis**
  - After incubation, centrifuge the samples at 12,000 rpm for 15 min to pellet the precipitated EPS.
  - Carefully discard the supernatant and collect the EPS pellets.
  - Store the EPS pellets at 4 °C or proceed with further analysis (*e.g.* quantification using the phenol-sulfuric acid assay or functional assays).

12.4 Quantification of EPS After Storage

12.4.1 Thawing and Preparations

- **Thawing of EPS samples:** Remove the stored EPS samples from the freezer and allow them to thaw at room temperature. If necessary, mix the sample carefully after thawing to ensure a homogeneous distribution of the polysaccharides. For highly viscous samples, heating the tube in a water bath (approximately 30–37 °C) for a few minutes may help to dissolve precipitated EPS.
- **Centrifuge (if necessary):** If the sample is still very viscous or contains particulate matter, centrifuge at 5000×*g* for 10 min at room temperature to clarify the sample. Carefully transfer the supernatant to a clean tube for further analysis.
  - **Note:** If EPS forms a gel-like structure upon thawing, you may need to dilute the sample with distilled water or a suitable buffer (*e.g.* PBS) to reduce the viscosity.

- **Preparation of standards:** Prepare a series of carbohydrate standard solutions with a known sugar such as glucose or mannitol. Prepare the standards in the range of 10–100 µg/mL (or other appropriate concentrations based on the expected EPS concentration in your samples).

12.4.2 Phenol-Sulfuric Acid Assay

- **Preparation of the reagents**
  - **Phenol solution (80%)**: Dilute concentrated phenol to 80% with distilled water.
  - **Concentrated sulfuric acid (H_2SO_4)**: Fresh and safely handled concentrated sulfuric acid.
- **Reaction setup**
  - Pipette 100 µL of each EPS sample into separate microcentrifuge tubes. If the sample is highly concentrated or viscous, use a suitable (*e.g.* 1:10) to bring it into the measurable range.
  - Add 500 µL of an 80% phenol solution to each tube.
  - Immediately add 2.5 mL of concentrated sulfuric acid to each tube. Handle the sulfuric acid carefully in a fume hood, as it is highly corrosive and generates heat when mixed.
- **Incubation**
  - Mix the tubes well by vortexing them briefly.
  - Incubate the tubes in a boiling water bath or heating block at 100 °C for 5 min. The reaction between the phenol, the sulfuric acid, and the polysaccharides will produce a yellow-orange color.
- **Cooling and measurement**
  - Allow the tubes to cool to room temperature after incubation.
  - Measure the absorbance of each sample and the standard at 490 nm using a UV-Vis spectrophotometer. This will provide a colorimetric reading proportional to the concentration of polysaccharides in each sample.
- **Calculate the polysaccharide concentration**
  - Using the absorbance values of your standards, create a calibration curve by plotting the absorbance (y-axis) against the known concentrations of the standards (x-axis).
  - Calculate the EPS concentration from the absorbance of your samples by referring to the calibration curve.

$$\text{EPS Concentration}\left(\mu g / \text{mL}\right) = \left(\text{Absorbance of the sample} / \text{Slope of the calibration curve}\right) - \text{Intercept of the calibration curve}$$

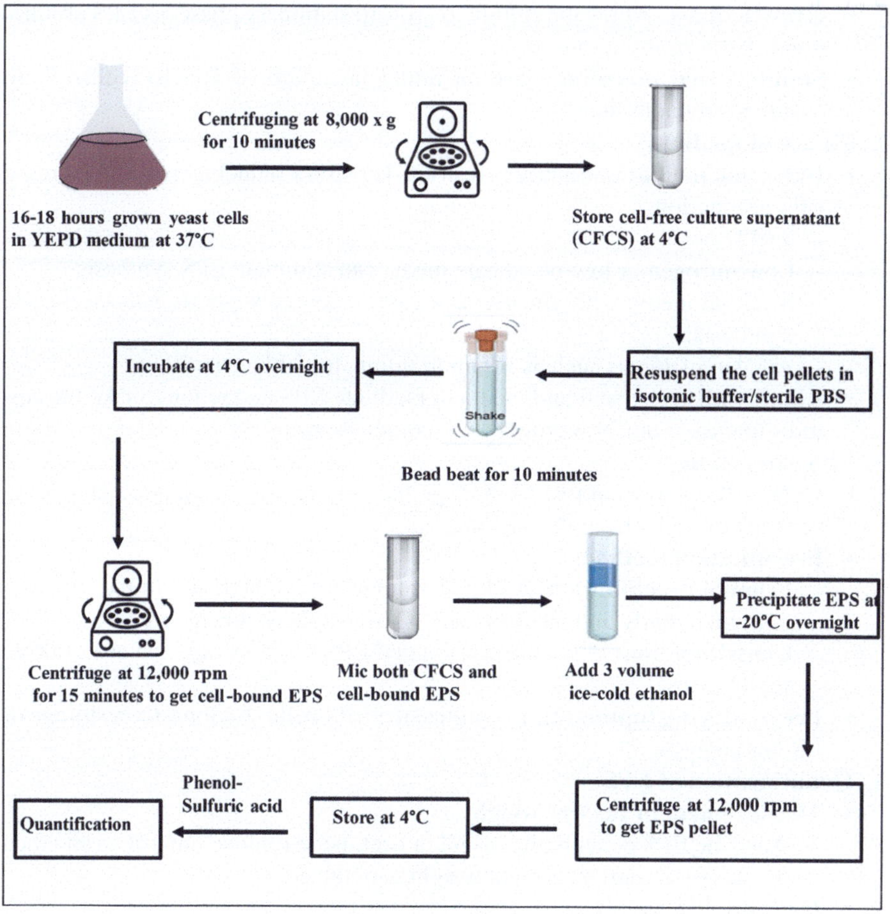

Fig. 12.1 Process for EPS extraction from fungal culture

- **Result**
  - The calculated EPS concentration can be expressed as μg EPS per mL supernatant. If you have measured the volume of the original fungal culture, you can also express the result as μg EPS per unit volume of culture (Fig. 12.1).

12.5 Tips and Troubleshooting

12.5.1 Tips

1. **Preparation of fungal culture**
   - **Fungal strain:** Use a pathogenic fungal strain known to secrete EPS (*e.g. Cryptococcus neoformans* for total glucan content, glucuronoxylomannan (GXM)/galactoxylomannan (GalXM) or *Candida albicans* for β-glucan).

- **Growth phase:** Allow the culture to grow to stationary phase, as EPS production usually peaks at this stage.
- **Standardized inoculum:** Use an initial inoculum of 10^6–10^7 cells/mL to ensure reproducibility.

2. **Choice of medium**
   - **Induction media:** Use nutrient-restricted or stress-inducing media to increase EPS production:
     - YEPD or
     - Low-nitrogen or low-phosphate media can stimulate EPS synthesis.
     - Minimal media with glucose as a carbon source supports polysaccharide synthesis.
     - Add supplements such as serum to mimic host-like conditions.
   - **Liquid culture:** Use liquid media to facilitate EPS extraction. Shake the cultures to ensure uniform growth and oxygenation.
3. **EPS extraction**
   - **Collect the supernatant:** Centrifuge the culture after incubation at 4 °C to separate the cells from the supernatant.
   - **Precipitation methods**
     - **Ethanol precipitation:** Add 2–3 volumes of cold ethanol to the supernatant, mix gently and incubate at 4 °C overnight to precipitate EPS.
     - **Centrifugation:** Pellet the precipitated EPS by centrifugation at 12,000×*g* for 15 min.
   - **Freeze-drying (optional):** Lyophilize the EPS pellet for long-term storage or further analysis.
4. **Quantification of EPS**
   - **Measurement of the dry weight**
     - Wash the EPS pellet with ethanol or acetone to remove impurities and then dry at 60 °C until a constant weight is obtained.
   - **Colorimetric assays**
     - **Phenol-sulfuric acid assay:** Measures total carbohydrate content. Use glucose or mannose as a standard.

12.5.2 Troubleshooting

1. **Problem: Low EPS yield**
   - **Possible causes:**
     - Non-inductive media or suboptimal growth conditions.
     - Insufficient incubation time.
   - **Solutions:**
     - Optimize the composition of the medium, paying attention to the carbon/nitrogen ratios.
     - Extend the incubation time but watch out for contamination or cell death.
     - Increase the initial inoculum density for higher biomass production.

2. **Problem: Contaminant in EPS**
   - **Possible causes:**
     - Carryover of proteins, DNA or small molecules during extraction.
   - **Solutions:**
     - Treat the supernatant with DNase and RNase to remove nucleic acids.
     - Use protease treatment to degrade proteins prior to ethanol precipitation.
     - Dialyze the EPS solution thoroughly with distilled water.
3. **Problem: Precipitation failure**
   - **Possible causes:**
     - Too little ethanol or improper mixing during precipitation.
   - **Solutions:**
     - Ensure that the ethanol is ice-cold for efficient precipitation.
     - Use 2–3 volumes of ethanol based on the volume of supernatant volume.
     - Allow sufficient incubation time (overnight at 20 °C).
4. **Problem: Difficulty in resuspending EPS**
   - **Possible causes:**
     - Overdried or aggregated EPS pellet.
   - **Solutions:**
     - Resuspend in warm distilled water or buffer with gentle vortexing.
     - Add a small amount of NaCl or dilute alkali solution (*e.g.* 0.1 N NaOH) if EPS aggregation persists.
5. **Problem: Poor reproducibility**
   - **Possible causes:**
     - Variability in inoculum size or culture conditions.
   - **Solutions:**
     - Repeat experiments under identical conditions.
     - Monitor pH, temperature, and aeration carefully.

Biofilm Formation Test and Determination of the Effect of Drugs on Their Formation

13

Abstract

Fungi form biofilms by attaching to surfaces and forming a protective matrix of extracellular polymeric substances (EPS), resulting in resilient microbial communities that can withstand environmental stresses and evade host immune system attacks (Malinovská et al, J Fungi 9:955, 2023). EPS comprises macro- and micromolecules that perform numerous functions, including adhesion, cell aggregation, protection against antimicrobial agents and host defense, provision of nutrients, stabilization of cell communities, and facilitation of genetic exchange. Different biology, environmental conditions, physical stress, and nutrient availability cause cells to secrete different components and proportions of EPS, leading to significant differences in the composition, three-dimensional structure, and functionality of the biofilm substrates formed. The protective nature of EPS makes combating biofilm-associated infections a particular challenge. In addition, EPS production enhances adhesion to the cell surface and promotes the retention of certain heavy metals, further complicating treatment strategies (Wang et al, Heliyon 10:e32766, 2024). Biofilms play a crucial role in the pathogenesis of fungal infections as they increase fungal resistance to antifungal treatments and contribute to persistent infections. Biofilms play a crucial role in various applications, including biotechnology, where they can be used for bioremediation and wastewater treatment. Understanding biofilm dynamics is crucial for developing effective antifungal therapies, as biofilm-associated infections often resist conventional treatments (Ali et al, Microorganisms 11:1934, 2023). In this chapter, basic protocols for the formation of fungal biofilms in a 12-well plate are described. The fungi are cultured in special media to promote biofilm formation. The process begins with the attachment of the fungi to the surface of the well, followed by the secretion of EPS to form the biofilm matrix. In subsequent steps, the biofilm must be maintained under favorable growth conditions, which typically includes optimizing temperature and nutrients to ensure

R. Pasrija, D. Kumari, *Protocols for Studying Pathogenic Fungi*,
https://doi.org/10.1007/978-981-96-7047-5_13

robust biofilm development. To study the effects of antifungal drugs on biofilm formation, the biofilm is exposed to the drug at different concentrations. Treatment with the drug aims to disrupt the architecture of the biofilm and prevent further development. The disruption of the biofilm is quantified using crystal violet staining, a widely recognized method for measuring the biomass of the biofilm. Crystal violet binds to the EPS and cellular components of the biofilm, allowing visualization and quantification of biofilm formation. After staining, biofilm biomass is assessed by measuring the absorbance of the dye, providing valuable insight into the effectiveness of the antifungal treatment in disrupting biofilm formation. The data obtained can provide important information on how different drugs affect the structure of the biofilm and its resistance to antifungal treatments. These findings are crucial for a better understanding of fungal biofilms and for the development of more effective strategies to combat fungal infections caused by biofilms. This chapter, makes an important contribution to mycology and antifungal therapy by providing a detailed protocol for biofilm formation and disruption. The knowledge gained from such studies is essential to improve the treatment of fungal infections and to develop new therapeutic approaches to effectively combat biofilms.

Keywords

Biofilms · Microbial communities · Infections · Drugs · Host defense · Protocol

Introduction This chapter covers detailed protocols ranging from biofilm formation to quantification. The following topics are covered as follows:

13.1 Materials Required
13.2 Preparations
13.3 Biofilm Formation and Disruption
13.4 Quantification of Biofilms by Crystal Violet Staining and Microscopy
13.5 Tips and Troubleshooting

13.1 Materials Required

13.1.1 Fungal Strains and Growth Media

- **Fungal strains:** Pure cultures of biofilm-forming fungi (*e.g. Candida* sp./*Cryptococcus* sp.).
- **Growth media**
  - **Roswell Park Memorial Institute (RPMI)-1640**: Frequently used for biofilm formation studies and often buffered with MOPS.

- **Yeast extract peptone dextrose (YEPD)** or **Sabouraud Dextrose Broth (SDB)**: For fungal precultures or biofilm formation.
- **Spider medium/Minimal medium (MM)** (optional)
- **Biofilm-inducing media**: Add 10% fetal bovine serum (FBS) to promote biofilm formation in some species.

13.1.2 Materials for Cultivation and Assay Setup

- **Sterile 12-well plates (flat bottom):** For biofilm formation assays.
- **Sterile tubes (*e.g.* 15 mL or 50 mL):** To prepare the fungal inoculum.
- **Pipettes and sterile tips:** For accurate liquid handling of liquids.
- **Centrifuge:** For pelleting the fungal cells when preparing the inoculum.
- **Sterile water or phosphate-buffered saline (PBS):** For washing biofilms.

13.1.3 Materials for Drug Testing

- **Antifungal drugs:** Amphotericin B, fluconazole, caspofungin, *etc.*, depending on the study. Prepare drug stock solutions in suitable solvents (*e.g.*, dimethyl sulfoxide (DMSO) or water).
- **Solvents:** DMSO (for dissolving hydrophobic drugs) and sterile water (for water-soluble drugs).

13.1.4 Biofilm Staining and Quantification

- **Crystal violet staining (0.2%):** For quantification of total biofilm biomass.
- **Ethanol (95% or absolute):** To dissolve bound crystal violet for spectrophotometric measurement.
- **Fluorescent dyes (optional):** Live/dead stain (*e.g.* propidium iodide) (to differentiate live and dead cells in biofilms). Calcofluor white can also be used to stain fungal biofilms.

13.1.5 Equipment for Detection

- **Microplate reader**: For absorption (*e.g.* crystal violet assay) and fluorescence (*e.g.* live/dead assay).
- **Light microscope or phase contrast microscope (optional)**: For observation of biofilm formation on slides or microtiter plates.
- **Confocal laser scanning microscope (CLSM):** For 3D visualization of biofilms and drug effects.

13.1.6 Controls (Optional)

- **Positive control**: Untreated biofilm-forming fungal culture.
- **Negative control:** Media without fungal inoculum or with non-biofilm-forming strains.
- **Drug vehicle control:** To consider the effects of solvents such as DMSO.

13.2 Preparations

13.2.1 Preparation of YEPD Media (See Chap. 2)

13.2.2 Preparation of RPMI-1640 (See Chap. 2)

13.2.3 Preparation of PBS (See Chap. 3)

13.2.4 Preparation of Saline Solution (0.85%) (See Chap. 3)

13.3 Formation of Biofilms and Disruption

13.3.1 Ingredients

- YEPD media, RPMI-1640, 12-well plate, 0.85% saline solution, PBS, coverslips, forceps and drug stock, *etc.*

13.3.2 Procedure

- **Preparation of the yeast suspension**
  - Grow the yeast cells overnight in YEPD media at 30 °C and shake at 150 rpm for 24 h.
  - After incubation, centrifuge the culture at 3000×*g* for 5–10 min to collect the cells.
  - Wash the pellet twice with PBS to remove residual media or contaminants.
  - Resuspend the cells in RPMI-1640 to an OD_{600} of 0.1.
- **Insert the coverslips into the 12-well plate**
  - Sterilize the coverslips (preferably round coverslips for optimal fit) by autoclaving or using a chemical sterilization method.
  - Using sterile forceps, place a sterilized coverslip in each well of a sterile 12-well plate.

- **Inoculation for biofilm formation**
  - For each strain, add 2 mL of RPMI-1640 yeast suspension (OD_{600} = 0.1) to each well of the 12-well plate with the sterilized coverslips.
  - Incubate the plate at 30 °C without shaking for 6 h (initial adhesion phase) to allow the yeast cells to adhere to the surface of the coverslips.
- **Maturation of the biofilm**
  - After the adhesion phase, remove the RPMI-1640 medium and replace it with 2 mL of fresh RPMI medium in each well.
  - Incubate the plate at 30 °C and shake at 150 rpm for 48 h to allow the biofilm to mature.
- **Wash to remove non-adherent cells**
  - After 48 h of incubation, gently wash the wells three times with PBS to remove any non-adherent or planktonic yeast cells.
  - The cells that remain attached to the surface of the coverslip are considered part of the biofilm.
- **Controls for biofilm formation and sterility**
  - Include well of the untreated yeast cells (without drug) as a control for biofilm formation.
  - Include wells with RPMI-1640 medium alone as a sterility control for the medium.
- **Evaluation of the anti-biofilm activity**
  - After the 6-h initial adhesion phase, add 2 mL of RPMI-1640 containing drug (normally at concentrations below the minimum inhibitory concentration (MIC)) to each well (see Chap. 5 for the MIC test procedure).
  - Incubate the plates at 30 °C for 48 h and shake at 150 rpm to evaluate anti-biofilm activity.
  - **Note:** The choice of media for biofilm formation varies depending on the fungal strain. While YEPD and SDB are commonly used for overnight fungal growth, other media such as RPMI-1640, Dulbecco's Modified Eagle Medium (DMEM), Spider Medium and various minimal media can also support biofilm formation. It is important that you adapt the incubation time and temperature to the respective fungal species. For example, *Candida* sp. usually requires 30 °C for 48 h, while *Cryptococcus* sp. requires 37 °C for 72 h [2].

13.4 Quantification of Biofilms after Crystal Violet Staining and Microscopy

13.4.1 Materials Required

- Crystal violet dye, 95% ethanol, spectrophotometer, cuvettes and light microscope, *etc.*

13.4.2 Procedure

- **Crystal violet staining:** Add 1 mL of 0.2% crystal violet solution to each well of the 12-well plate with the biofilm-covered coverslips. Incubate the plate at room temperature for 10 min to allow the crystal violet dye to bind to the biofilm.
- **Wash to remove excess dye:** After the staining time, wash each well thoroughly three times with sterile water or PBS to remove the unbound dye and ensure that only the dye bound to the biofilm remains.
- **Microscopic observation:** Place the coverslips stained with crystal violet on a clean microscope slide to visualize the biofilm structure. Examine the biofilm structure under a light microscope with a 40× objective lens. Compare the biofilms of the treated and control groups to observe differences in biofilm formation and structure (as shown in Fig. 13.1).
- **Decolorize the biofilms:** After washing, add 200 μL of 95% ethanol to each well. Incubate at room temperature for 20 min to decolorize the biofilm to allow the crystal violet dye bound to the biofilm to dissolve in the ethanol solution.
- **Measure the absorbance (OD_{570}nm):** Transfer the decolorized solution from each well to a cuvette. Measure the OD at 570 nm using a spectrophotometer. Perform the measurements in triplicates for each condition to ensure reproducibility.
- **Calculation of biofilm inhibition:** Calculate the percentage inhibition of biofilm formation in the presence of the antimicrobial agents using the following formula:

$$100-\left[\left(\text{treated cells O.D.}\times 100\right)/\text{untreated cells O.D.}\right].$$

This formula compares the optical density of biofilms treated with antimicrobial agents (treated cells) to those without treatment (untreated cells) and provides a percentage that reflects biofilm inhibition [1].

13.5 Tips and Troubleshooting

13.5.1 Tips

1. **Fungal strain and preparation of the inoculum**
   - **Use known biofilm-forming strains:** Work with species such as *C. albicans* SC5314 or *C. neoformans* for reproducible biofilm formation.
   - **Standardized inoculum density:** Use a cell density of 10^6–10^7 cells/mL to achieve uniform biofilm formation.
     - Prepare cells from the mid-logarithmic growth phase to achieve higher viability and biofilm formation potential.
   - **Wash the cells:** Wash fungal cells in PBS or saline to remove planktonic growth and metabolic by-products before inoculation.

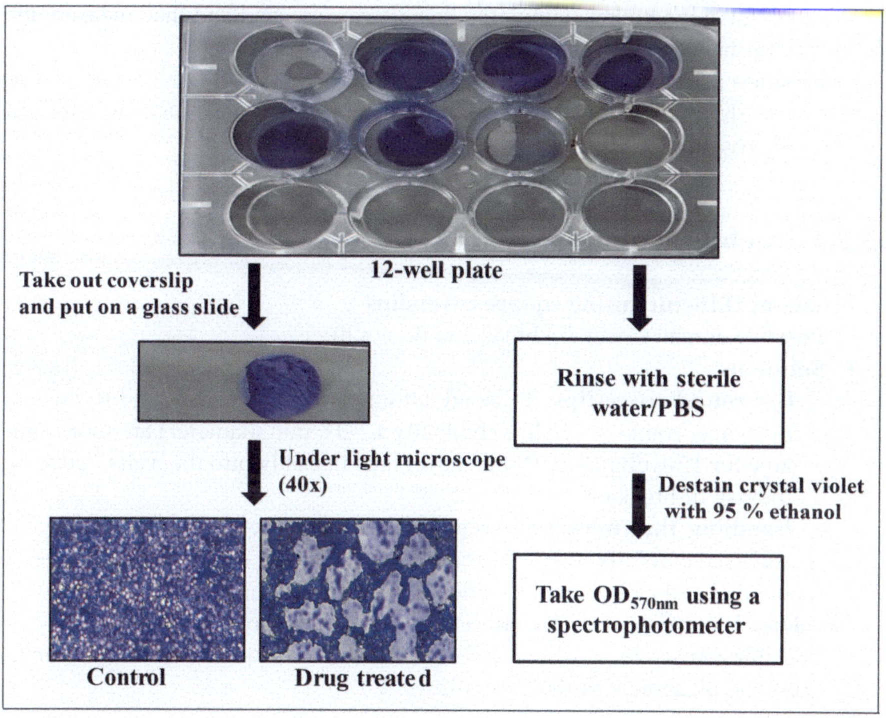

Fig. 13.1 Image of the 12-well plate with biofilms on the coverslip stained with crystal violet and biofilms under the microscope to quantify the fungal biofilms

2. **Choosing the right surface**
   - Use 12-well plates or glass coverslips, as these promote biofilm adhesion.
   - Test alternative surfaces (*e.g.* silicone, stainless steel) when examining biofilms under device-like conditions.
3. **Selection of media**
   - **Standard media:** Use RPMI-1640 (with or without serum), YEPD, or Spider medium.
   - **Nutrient limitation:** Media with limited nutrients (*e.g.* minimal media) can promote biofilm formation under stress.
   - **pH and supplements:** Adjust pH (typically 7.0) and consider adding 10% FBS for biofilm-inducing conditions.
4. **Biofilm Incubation**
   - **Temperature and time:** Incubate at optimal conditions (*e.g.* 30 °C and 48 h for *Candida* sp./37 °C and 72 h for *Cryptococcus* sp).
   - **Static conditions:** Avoid shaking for the first 4–6 h to allow cell adhesion.
5. **Quantification of the biofilm**
   - **Crystal violet assay**
     - Stain biofilms with 0.2% crystal violet for 15–30 min.
     - Wash thoroughly with water to remove excess stains.

  - Dissolve the bound stain with ethanol or acetic acid and then measure the absorbance at 590 nm.
- **Biomass and thickness (optional)**
  - Use dry weight measurement or imaging techniques such as confocal microscopy or SEM.

13.5.2 Troubleshooting

1. **Problem: Difficulty using square coverslips**
   - **Possible cause**: Coverslip break and do not fit.
   - **Solutions**:
     - **Use round coverslips**: To avoid cutting square coverslips and to prevent breakage, round coverslips (typically 12–18 mm diameter) are more suitable for 12-well plates. Round coverslips fit easily into the wells, reducing the risk of breakage.
     - **Handling the tweezers/forceps**: Ensure the tweezers/forceps are sterile and used carefully to avoid excessive force, which can lead to breakage. A steady hand and gentle placement can minimize the risk.
2. **Problem: Inconsistent or incomplete biofilm formation**
   - **Possible cause**: Inconsistent cell inoculation, inadequate incubation conditions and placement of the coverslip.
   - **Solutions:**
     - **Ensure uniform inoculation**: Ensure the yeast cells are uniformly suspended in RPMI-1640 medium at the correct OD_{600} (0.1) before adding to wells. This ensures uniform inoculation in all wells.
     - **Optimal incubation conditions**: See point 4 of Sect.13.5.1 for optimal incubation conditions.
     - **Stability of the coverslips**: When inserting the coverslips, make sure that they are carefully placed in the wells. You should check the wells regularly to ensure the coverslips do not slip during incubation.
3. **Problem: Insufficient or inconsistent biofilm staining**
   - **Possible causes:**
     - **Crystal violet staining**: Inconsistent or incorrect staining may be caused by improper washing of wells, poor dye concentration, or insufficient incubation time.
     - **Washing**: Incomplete or inconsistent washing of wells may result in excess crystal violet, which will affect the quantification or appearance of biofilms.
   - **Solutions**:
     - **Proper washing**: Ensure wells are thoroughly washed after the biofilm incubation phase. Wash three times with PBS or sterile water to remove non-adherent cells before staining.
     - **Optimal dye concentration and incubation time**: Use the correct concentration of 0.2% crystal violet and incubate for 10–30 min at room tem-

perature. Ensure that the dye has sufficient time to bind to the biofilm. If little biofilm forms, it may be helpful to slightly increase the concentration of crystal violet.
   - **Ensure proper drying**: After staining with crystal violet, allow the biofilm on the coverslips to dry well (2–3 h) before microscopy to ensure a clear view.
   - **Decolorize**: After crystal violet incubation, wash thoroughly with sterile water or PBS to remove unbound dye. When decolorizing with ethanol, ensure that the incubation time (20 min) is sufficient to remove excess dye and keep the biofilm visible. If necessary, stain for longer (up to 30 min).

4. **Problem: Poor or inconsistent anti-biofilm activity during drug treatment**
   - **Possible causes:**
     - **Incorrect drug concentration**: using sub-MIC concentrations of the drug is critical for evaluating biofilm disruption. If the concentration of the drug is too high, it may kill the yeast cells directly, instead of specifically destroying the biofilm matrix.
     - **Inconsistent exposure**: The biofilm may not be exposed to the drug for 48 h, or the drug may not be well mixed into the RPMI-1640 medium.
     - **Drug Interaction**: The choice of drug or its interaction with the medium may interfere with biofilm formation or the fungal growth itself, leading to misleading results.
   - **Solutions:**
     - **Validate the drug concentrations**: Ensure that drug concentrations are calculated correctly and are the same in all wells. Use appropriate sub-MIC concentrations (usually lower than the MIC but sufficient to assess biofilm disruption).
     - **Proper drug addition and incubation**: Add the drug after the adhesion phase (6 h) to ensure that the drug is incorporated into mature biofilms. For optimal incubation conditions, see point 4 in Sect. 13.5.1. Check that the drug is evenly distributed in the RPMI-1640 medium.
     - **Include controls**: Always include positive and negative controls in your assay to confirm that the biofilm formation and dissolution results are due to the drug's effect and not to variations in the experimental setup.

5. **Problem: Microscopy problems (biofilm visualization)**
   - **Possible causes:**
     - **Low-resolution images**: The biofilm structure might be difficult to see at low magnification or poor resolution.
     - **Inadequate sample preparation**: Improper preparation of stained biofilms on coverslips may result in poor visualization.
   - **Solutions:**
     - **Use appropriate magnification**: Use 40× magnification or higher for visualization preferably with an oil immersion lens for better clarity. Ensure that the image resolution is high enough to capture detailed biofilm structures.

– **Sample preparation for microscopy**: When preparing the stained coverslips, ensure they are thoroughly dried and clean. Place them carefully on the slide to avoid artifacts that obscure the biofilm structure.

References

1. Márton R, Nagy B, Molnár M (2023) Biofilm development of *Candida boidinii* and the effect of tyrosol on biofilm formation. Biotechnol Lett 45(11):1541–1554. https://doi.org/10.1007/s10529-023-03432-5
2. Verstrepen KJ, Klis FM (2006) Flocculation, adhesion and biofilm formation in yeasts. Mol Microbiol 60(1):5–15. https://doi.org/10.1111/j.1365-2958.2006.05072.x

Induction of Capsule Formation in *Cryptococcus* sp.

14

Abstract

Cryptococcus neoformans is a pathogenic yeast characterized by the production of a conspicuous polysaccharide capsule, which is critical for its virulence and ability to evade the host immune response. The capsule, a complex structure composed of 88% glucuronoxylomannan (GXM) and 12% galactoxylomannan (GalXM) surrounds the fungal cell and plays an important role in pathogenicity. A special feature of the *C. neoformans* capsule is its hydrophilicity and the fact that its refractive index is very close to that of the surrounding medium, which makes it invisible under the light microscope without special techniques. Capsule formation is a dynamic process influenced by growth conditions and can be visualized using negative staining methods, the most commonly used method being India ink staining. In this protocol, we describe the standard techniques for triggering capsule formation and subsequent visualization of the capsule using India ink negative staining in two different growth media: Dulbecco's Modified Eagle Medium (DMEM) and Sabouraud Dextrose Broth (SDB). DMEM is often used for co-culture experiments with mammalian cells and provides conditions more similar to those in a host environment. SDB, on the other hand, is often used for fungal cultures and supports the robust growth of *C. neoformans*. Capsules can be effectively induced in both DMEM and SDB, although size and structure may vary depending on the medium used. After incubation, India ink is added to the fungal suspension to stain the background while the capsule remains unstained. This creates a clear "halo" around each cell, which is characteristic of capsule formation (Gutierrez-Gongora et al, Sci Rep 13:4928, 2023). This method allows a qualitative analysis of the capsule size, shape and overall capsule density of the fungal cells. The capsule is a key factor in studies of virulence, immune evasion and pathogenesis, and this protocol serves as an important tool to visualize and quantify the capsule in laboratory settings. This method allows researchers to study the influence of different environmental conditions, antifungal treatments

R. Pasrija, D. Kumari, *Protocols for Studying Pathogenic Fungi*,
https://doi.org/10.1007/978-981-96-7047-5_14

and genetic factors on the production of the *C. neoformans* capsule, which is an important aspect of its pathogenic potential (Zaragoza et al, Adv Appl Microbiol 68:133–216, 2009).

Keywords

Cryptococcus · Capsule · India ink · Translucent area · Staining

Introduction This chapter provides detailed protocols ranging from capsule formation to size calculation. Topics covered include:

14.1. Materials Required
14.2. Preparations of Media and Buffers
14.3. Protocol for Capsule Formation and Analysis
14.4. Tips and Troubleshooting

14.1 Materials Required

14.1.1 Fungal Strains and Growth Media

- **Fungal strains:** Pure culture of *Cryptococcus* sp., especially capsule-producing strains such as *Cryptococcus neoformans* or *Cryptococcus gattii*.
- **Growth media**
  - **Sabouraud dextrose broth (SDB)**: For precultures or general fungal growth.
  - **Capsule-inducing media:** Dulbecco's Modified Eagle Medium (DMEM) and serum-saturated media (*e.g.* Fetal Bovine Serum (FBS)).

14.1.2 Environmental Conditions

- **Settings for the incubation**
  - High CO_2 concentration (5–10%) in a CO_2 incubator for optimal capsule induction.
  - Standard temperature: 30/37 °C, depending on strain and experimental conditions.
  - **Note:** For *Cryptococcus* sp. a stronger capsule induction at 37 °C is reported

14.1.3 Reagents for Visualization of the Capsules

- **India ink or Nigrosin stain:** To produce a negative contrast for observation of the capsule under the light microscope.

14.1.4 Microscopy and Imaging Techniques

- **Light microscope**: Phase-contrast or bright-field microscope for observation of the capsule with India ink staining.
- **Confocal laser scanning microscope (CLSM) (optional)**: For detailed 3D visualization of the capsules.

14.1.5 Materials for Handling the Cultures

- **Sterile Petri dishes or culture flasks:** For cultivating fungal cells in capsule-inducing media.
- **Sterile centrifuge tubes (*e.g.* 15 mL or 50 mL):** For harvesting fungal cells after incubation.
- **Centrifuge:** To pellet the fungal cells after culture (3000–5000 rpm for 5–10 min).
- **Sterile pipettes and tips:** For inoculation and media handling.
- **Sterile water or Phosphate Buffered Saline (PBS):** For washing the fungal cells.

14.1.6 Materials for the Quantification of Capsule Formation

- **Colorimetric test reagents (optional)**: Phenol-sulfuric acid reagent (to quantify the total polysaccharides released in the medium).
- **Size measurement tools**: Microscope (for manual measurement of capsule thickness).
- **Image analysis software (optional):** For analyzing the capsule size

14.1.7 Controls (Optional)

- **Positive control:** (conditions known to trigger capsule formation).
- **Negative control:** Non-capsule triggering media to verify the absence of capsule formation.

14.1.8 Safety and Aseptic Materials

- **Personal protective equipment**: Lab coat, gloves, and goggles to ensure biological safety.
- **Bunsen burner or alcohol lamp:** For aseptic handling of cultures.
- **Autoclave or sterilizer:** For the sterilization of media, tools, and waste.

14.2 Preparations of Media and Buffers

14.2.1 Preparation of YEPD Media (See Chap. 2)

14.2.2 Preparation of SDB Media

- **Materials required:** Dextrose (glucose) (40 g), peptone (10 g), agar (optional, if preparing a solid medium) (15 g), distilled or deionized water, pH meter or pH paper (for pH adjustment), autoclave or sterilization equipment, glass bottles or flasks (for media preparation), magnetic stirrer (optional)
- **Procedure**
  - **Weigh the ingredients:** Weigh out 40 g of dextrose (glucose) and 10 g of peptone
  - **Dissolve the components:** Add the 40 g dextrose and 10 g peptone in a 1 L glass bottle or glass flask. Add about 900 mL of distilled or deionized water to the flask.
  - **Stir the mixture:** Gently stir the solution with a magnetic stirrer (or by hand) until all components are completely dissolved.
  - **Adjust the final volume:** Once the ingredients are completely dissolved, adjust the final volume to 1 L by adding distilled or deionized water to the flask.
  - **Check the pH value:** Measure the pH value of the medium with a pH meter. The pH value should be approximately 5.6. If necessary, adjust the pH to 5.6 with 1 M NaOH (to increase the pH) or 1 M HCl (to decrease the pH).
  - **Sterilize the media:** Autoclave the media at 121 °C for 15–20 min to sterilize them. If working under non-sterile conditions, ensure the equipment is clean and contamination is minimized.
  - **Storage:** Allow the media to cool to room temperature after autoclaving. The prepared SDB can be stored at room temperature.

14.2.3 Preparation of PBS (See Chap. 3)

14.3 Protocol for Capsule Formation and Analysis [1–3]

14.3.1 Ingredients

- YEPD, DMEM/SDB, PBS, fungal culture, spectrophotometer, incubator, India ink, glass slides, coverslips, microscope, *etc.*

14.3.2 Procedure

- **Inoculate and grow the yeast:** Inoculate *C. neoformans* from a single colony in YEPD medium and incubate overnight at 30 °C and 150 rpm for 24 h. This promotes optimal yeast growth and metabolic activity.
- **Harvest the yeast cells:** Centrifuge the culture after 24 h of incubation at 3000×*g* for 5–10 min to pellet the yeast cells. Carefully decant the supernatant (the spent medium) and discard it.
- **Wash the yeast cells and resuspend them in DMEM/SDB medium:** Resuspend the yeast pellet in PBS to remove any residual growth medium and non-specific components. Centrifuge again at 3000×*g* for 5 min to pellet the cells. Discard the supernatant and repeat the washing step twice to ensure that the yeast cells are thoroughly cleaned.
- **Resuspend in capsule induction medium (DMEM/SDB):** After the second wash step, resuspend the yeast cells in a mixture of DMEM/SDB (diluted tenfold). The final suspension should have a concentration of about 1×10^7 cells/mL. Use sterile techniques when preparing and handling the suspension to avoid contamination.
- **Incubate the suspension:** Incubate the cell suspension at 30 °C/37 °C for 3 days. Keep the cells at a constant temperature to allow optimal capsule formation. During this incubation period, *C. neoformans* will form a conspicuous polysaccharide capsule that can be visualized after the appropriate staining.
- **Prepare the India ink stain:** After 3 days of incubation, take a small aliquot of the yeast suspension (approximately 10–20 μL). Mix this aliquot with India ink at a ratio of 1:1. The India ink will stain the background while the capsule (if present) remains unstained, creating a clear halo around each yeast cell.
- **Prepare the slide:** Place a drop of the India ink and yeast mixture in the center of a clean slide. Use a second slide or a coverslip to spread the mixture evenly and create a thin smear. If you want a thicker smear, you can also leave the sample undisturbed.
- **Examine the sample under a microscope:** Allow the smear to air dry for a few minutes to ensure that the sample adheres to the slide and excess ink evaporates. Observe the smear under a light microscope at 100× magnification and use the oil immersion lens for the best resolution. The capsule appears as a clear halo around the cell (Fig. 14.1), contrasting with the darkly colored background. The yeast cells appear dark due to the India ink staining, while the surrounding capsule remains transparent, revealing its size and structure.

14.3.3 Quantification of Capsule Production

- For quantitative analysis, measure the cell size and capsule size using the images taken under the microscope.
  - **Cell size without capsule**: Measure the diameter of the yeast cell without the capsule.

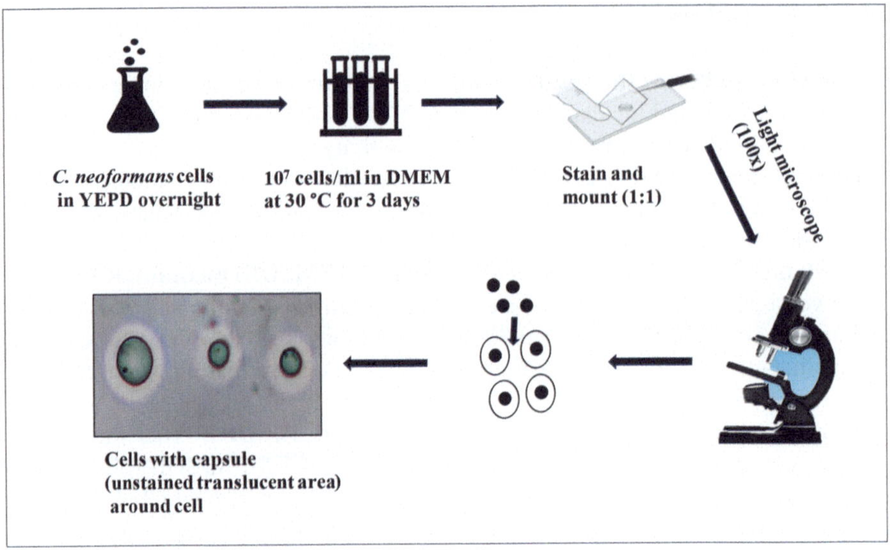

Fig. 14.1 Schematic representation of the capsule induction and microscopy procedure in *C. neoformans*

- **Total cell size with capsule**: Measure the total diameter of the yeast cell, including the capsule.
- The capsule size is determined by the difference between the total cell size (with capsule) and the cell size (without capsule).

• Perform the measurements for at least three biological replicates (*i.e.* separate cultures of *C. neoformans*) to ensure the reliability and reproducibility of the results.

• Use image analysis software (*e.g.* ImageJ or Adobe Photoshop PS3) to consistently quantify capsule size as described below:

- **Open and prepare the image in Adobe Photoshop CS3**
  - **Import the image:** open the captured image in Adobe Photoshop CS3 by selecting "File" > "Open" and navigate to the location where the image is stored.
  - **Convert to grayscale (if necessary):** If your image is colored (*e.g.* stained with India ink), convert it to grayscale for easier analysis:
    Go to Image > Mode > Grayscale. If prompted to discard color information, select Discard.
  - **Adjust the Image Contrast (optional):** You may need to adjust the image contrast to make cell and capsule boundaries more visible:
    Go to Image > Adjustments > Brightness/Contrast, and adjust the sliders until the cell and capsule are visible.
  - **Zoom in:** Zoom in on the cells to ensure that the cell body and the surrounding capsule are visible and well-delineated.

- **Measure cell diameter without capsule (cell body only)**
  - **Select the cell region:** Use the Elliptical Marquee Tool (press M) to carefully outline the cell body of the *Cryptococcus* cell. Be sure to exclude the capsule (halo) from the selection.
  - **Measure the cell diameter:** With the cell body selected, go to Analyze > Histogram. The pixel intensity distribution for the selected area is displayed here. The diameter of the cell can be estimated based on the selection. To do this accurately, note the width of the cell at its largest point using the Ruler Tool (press I to select the Ruler Tool). Note the diameter of the fungal cell (without the capsule).
  - **Measure the total diameter (cell + capsule)**
    - **Select the entire cell and the capsule:** Use the Elliptical Marquee Tool again, but this time, select the entire cell, including the surrounding capsule (halo). Make sure to capture the outermost boundary of the capsule.
    - **Measure the total diameter**: With the total area (cell + capsule) selected, go to Analyze > Histogram again. Measure the total diameter by plotting the distance from one side of the cell (including the capsule) to the other using the Ruler Tool. Note the total diameter (cell + capsule).
- **Calculate the capsule size**
  - **Subtract the cell diameter from the total diameter**: To determine the capsule size, subtract the cell diameter (without capsule) from the total diameter (cell + capsule).

$$\text{Capsule size} = \text{Total diameter}\left(\text{cell} + \text{capsule}\right) - \text{Cell diameter}\left(\text{without capsule}\right).$$

  - **Calculation example**: If the total diameter (cell + capsule) = 10 μm and the cell diameter (without capsule) = 5 μm, the capsule size would be:

$$\text{Capsule size} = 10\,\mu m - 5\,\mu m = 5\,\mu m.$$

- **Analyze multiple cells for statistical accuracy**
  - **Measure multiple cells**: To ensure reliable results, measure the diameter without capsule and the total diameter (cell + capsule) for at least 10–20 cells. This will compensate for variations between individual cells and provide more accurate results.
  - Statistical analysis: Perform statistical analysis of capsule size data (*e.g.* mean ± standard deviation) to assess variability and differences in capsule production between biological replicates.
- **Data interpretation**
  - Compare capsule size under different conditions (*e.g.* different growth media or experimental treatments) to evaluate factors that may influence capsule formation in *C. neoformans*.
  - Consider using graphical representations such as bar graphs or scatter plots to show the differences in capsule production between experimental groups.

14.4 Tips and Troubleshooting

14.4.1 Tips

1. **Selection of the fungal strain**
   - Use a strain with a known ability to form capsules, such as *C. neoformans* H99.
   - Ensure the strain is viable and stored under optimal conditions before use.
2. **Preparation of the inoculum**
   - **Start culture:** Grow the cells in a rich medium (*e.g.* YEPD broth) overnight at 30 °C.
   - **Washing the cells:** Wash the cells thoroughly with sterile PBS to remove residual nutrients that may interfere with capsule induction.
   - **Standardized inoculum:** Prepare a cell suspension of 10^6–10^7 cells/mL to ensure reproducibility.
3. **Induction medium and conditions**
   - **Capsule-inducing medium**
     - Use minimal media such as tenfold diluted SDB or low glucose medium (1% glucose) to simulate low nutrient conditions.
     - Supplement the medium with 10% FBS to mimic host-like conditions.
     - CO_2-enriched conditions to promote capsule formation.
   - **Incubation conditions**
     - Incubate the cultures at 37 °C to mimic the conditions in the human host.
     - Use a humidified chamber or CO_2 incubator (5% CO_2) for optimal results (optional).
     - Allow capsule induction for 2–3 days, depending on the experimental setup.
4. **Visualization of the capsules**
   - **India ink staining**
     - Mix an aliquot of the culture with India ink on a glass slide.
     - Observe under a light microscope to visualize the halo representing the capsule.
   - **Quantification of the capsule size**
     - Measure the capsule width using a microscope and image editing software (*e.g.* ImageJ/Adobe Photoshop CS3).
     - Calculate the capsule-to-cell ratio for accurate comparisons.

14.4.2 Troubleshooting

1. **Problem: No capsule formation (no halo around yeast cells)**
   - **Possible causes:**
     - The *C. neoformans* strain may not be in the optimal growth state for capsule induction.
     - Insufficient incubation time or temperature.

  - Incorrect medium composition or lack of suitable nutrients in the capsule induction medium (DMEM/SDB).
  - Yeast cells may be stressed, damaged, or nonviable.
- **Solutions:**
  - **Check the viability of the culture**: Ensure that the *C. neoformans* strain is healthy and capable of capsule formation. Some strains are unable to form capsules under standard conditions. Confirm capsule formation with control strains or previous successful experiments.
  - **Ensure optimal incubation**: Incubate the culture at 30 °C for 72 h. Capsule formation requires at least 3 days for complete development of a polysaccharide capsule.
  - **Proper medium**: Ensure that the medium used for capsule induction is DMEM/SDB and is sterile. An incomplete or contaminated medium may hinder capsule formation.
  - **Check yeast cell density**: Ensure that the initial inoculum has a concentration of 1×10^7 cells/mL to avoid overcrowding or insufficient growth.

2. **Problem: Inconsistent capsule formation (some cells have capsules, others do not)**
   - **Possible causes:**
     - The inoculum may not be uniform, leading to different growth conditions.
     - Not all cells may be in the same growth phase, which may affect their ability to produce a capsule
     - Inconsistent incubation conditions (*e.g.* temperature fluctuations or contamination) can lead to uneven capsule formation.
   - **Solutions:**
     - **Ensure uniform inoculation**: carefully resuspend the yeast pellet to ensure even distribution of cells. Ensure the starting culture is evenly mixed before transferring to the capsule induction medium.
     - **Ensure uniform growth**: To minimize this variability, it is important to use a uniformly growing culture. Ensure that all cells are in the logarithmic growth phase during capsule induction.
     - **Harvest at the right time**: After inoculating the yeast into the growth medium, the cells are incubated appropriately (usually 24 h in YEPD medium) to reach the logarithmic phase. At this time, they are metabolically most active and capable of capsule formation.
     - **Monitor the incubation conditions**: Maintain a consistent temperature of 30 °C throughout the incubation period. Temperature fluctuations can stress the cells and impair capsule formation.
     - **Prevent contamination**: Always use sterile procedures when handling the culture, media and slides. Contamination can interfere with capsule formation or lead to inconsistent results.
3. **Problem: Inconsistent or poor staining with India ink**
   - **Possible causes:**
     - The India ink smear may not be evenly distributed on the slide, resulting in uneven staining.

  - The India ink concentration may be incorrect so that it is either too faint or too dark to visualize the capsule properly.
- **Solutions:**
  - **Optimize the ink concentration**: Ensure the India ink is mixed in a 1:1 ratio with the yeast suspension. If the coloration is too dark, dilute the ink further. If it is too faint, increase the ink concentration slightly.
  - **Improve the smear technique**: Spread the India ink and yeast mixture evenly on the slide to obtain a thin smear. A thick smear can lead to uneven staining and difficulty identifying the capsules.
  - **Check the slide preparation**: Allow the smear to air dry before viewing under the microscope. Excess ink or an unclean slide surface can affect clear capsule visualization of the capsules.

4. **Problem: Poor visualization of the capsules under the microscope**
   - **Possible causes:**
     - The microscope may not be set up correctly (*e.g.* incorrect magnification or lack of oil immersion).
     - Insufficient contrast between the capsule and the cell body.
     - Poor quality of the smear preparation or insufficient staining of the sample.
   - **Solutions:**
     - **Use an oil immersion objective lens**: Ensure the microscope is equipped with a 100× oil immersion lens to achieve the necessary magnification to observe the capsule. The capsules are usually 1–5 μm in size, so a higher magnification is essential.
     - **Adjust the contrast and brightness**: If the capsule is faint, adjust the contrast and brightness of the image using the microscope settings. You may also need to adjust the image settings using image analysis software.
     - **Ensure correct staining**: If the capsule is still not visible, check the staining procedure. The India ink should leave a clear halo around each yeast cell so that the capsule stands out from the darkly stained cell body. Adjust the ink concentration and ensure that the smear is thin and uniform.
5. **Problem: Capsule formation in the controls (unexpected capsule formation in the negative controls)**
   - **Possible causes:**
     - The negative controls (*e.g.* cultures in media without capsule-inducing factors) may have been exposed to conditions that inadvertently promote capsule formation, such as high humidity, temperature, or stress.
   - **Solutions:**
     - **Check the control conditions**: Ensure negative controls are not exposed to capsule-inducing media (DMEM/SDB). Use non-inducing media (such as YEPD medium without DMEM/SDB) for the negative control cultures.
     - **Pay attention to temperature and light**: Ensure that the negative controls are incubated under conditions that would not favour capsule formation. In *C. neoformans*, capsule formation is triggered by the capsule induction medium and specific environmental conditions. For proper control, avoid temperature fluctuations (ensure constant incubation at 30 °C) and exposure to light.

6. **Problem: Inaccurate measurement of capsule size**
   - **Possible causes:**
     - Improper image analysis or inaccurate boundary selection may result in incorrect capsule size measurements.
     - The image quality could be poor, making it difficult to distinguish the cell body from the capsule.
   - **Solutions:**
     - **Ensure high-resolution imaging**: Acquire images using a high-resolution camera with appropriate settings for the oil objective. This will help to obtain clear and precise measurements.
     - **Improve boundary selection**: When measuring capsule size using software (*e.g.* ImageJ or Adobe Photoshop), ensure that the boundaries between the cell body and capsule are clearly defined. Be precise when selecting boundaries for accurate measurements. Avoid including non-capsule structures such as debris or other particles.
     - **Adjust the image contrast**: If the capsule is too faint to see, adjust the brightness and contrast of the image to increase visibility. This can help the capsule appear better in the final image.
     - **Measure multiple cells**: Measure the size of capsules from at least 10–20 cells to account for variability in capsule size. This will give you a more accurate representation of capsule formation throughout the culture.
7. **Problem: Capsule formation is less pronounced than expected (smaller capsules)**
   - **Possible causes:**
     - The induction medium (DMEM/SDB) may not have been prepared properly or essential components for capsule induction may be missing.
     - The culture conditions may not have been optimal for capsule formation (*e.g.* temperature too low and incubation time too short).
   - **Solutions:**
     - **Ensure the correct composition of the medium**: Check the composition of the capsule induction medium (DMEM/SDB). Ensure that all components are correctly prepared and sterile and that the medium is not old or damaged.
     - **Extend the incubation time**: Incubate the yeast cells in DMEM/SDB for a longer period of time (up to 4–6 days if required) to promote optimal capsule formation.
     - **Optimize the temperature**: Keep the incubation temperature stable at 30 °C. Although 30 °C is usually optimal for *C. neoformans*, you should test slight temperature fluctuations to determine the best conditions for capsule formation in your specific strain.
     - **Adjust the inoculum density**: Ensure that the yeast suspension is not too dense or too sparse. 1×10^7 cells/mL is generally ideal to induce capsule formation. Too high a density may lead to cell stress, while too low a density may not provide enough interaction between cells and medium for optimal capsule production.

8. **Problem: Fungal growth is stunted or poor**
   - **Possible causes:**
     – The culture may not be viable, or it may have been exposed to stress during the process (*e.g.* improper handling, temperature fluctuations, or contamination).
   - **Solutions:**
     – **Confirm the viability of the culture**: Check the original yeast culture for sufficient growth in the YEPD medium. If there is evidence of contamination or poor growth, discard the culture and start with a new, healthy colony.
     – **Use sterile techniques**: Always use sterile techniques when preparing cultures, handling media, and transferring samples to avoid contamination.
     – **Ensure proper incubation conditions**: Ensure the culture is incubated at 30 °C/37 °C and in an appropriate medium (DMEM/SDB) for adequate growth and capsule formation.

References

1. Guess T, Lai H, Smith SE et al (2018) Size matters: measurement of capsule diameter in *Cryptococcus neoformans*. J Vis Exp 57171. https://doi.org/10.3791/57171
2. Gutierrez-Gongora D, Raouf-Alkadhimi F, Prosser RS, Geddes-McAlister J (2023) Differentiated extracts from freshwater and terrestrial mollusks inhibit virulence factor production in *Cryptococcus neoformans*. Sci Rep 13:4928. https://doi.org/10.1038/s41598-023-32140-3
3. Haynes BC, Skowyra ML, Spencer SJ et al (2011) Toward an integrated model of capsule regulation in *Cryptococcus neoformans*. PLoS Pathog 7:e1002411. https://doi.org/10.1371/journal.ppat.1002411

Melanin Formation in *Cryptococcus* sp.

15

Abstract

Cryptococcus is an opportunistic pathogenic fungus that plays a significant role in human infections, especially in immunocompromised individuals. A hallmark of *Cryptococcus* virulence is its ability to synthesize melanin, a pigment that contributes to its resistance to various host defense mechanisms, including oxidative stress, immune responses, and thermal stress (Baker et al, Biol Chem 298(1):101519. 2021). The synthesis of melanin in *Cryptococcus* is catalyzed by the enzyme laccase, which converts precursors such as L-3,4-dihydroxyphenylalanine (L-DOPA) into melanin (Lee et al, mBio 10. https://doi.org/10.1128/mbio.02267-19, 2019). Since *Cryptococcus* cannot synthesize L-DOPA itself, it requires the exogenous addition of L-DOPA to the growth medium to initiate melanin production (Eisenman et al, Virulence 2:329–336, 2011). This external supply of L-DOPA is crucial for promoting melanin biosynthesis and allows researchers to study the pigmentation process under controlled laboratory conditions. This melanin production plays a key role in the ability of *Cryptococcus* to survive and thrive in the host and protect against reactive oxygen species (ROS) and other environmental stresses (Kumari et al, Arch Microbiol 206:355, 2024). The presence and function of melanin in *Cryptococcus* are critical to understanding its pathogenicity, making it an important subject of research in fungal biology and infectious diseases (Chrissian et al. J Biol Chem 295(7):1815–1828, 2020). This chapter provides detailed protocols for studying melanin formation in *Cryptococcus* sp., including the preparation of minimal media (MM) for optimal melanin synthesis. The process of melanin formation is described, highlighting the biochemical pathways involved and emphasizing the role of laccase in the conversion of L-DOPA to melanin. This chapter also discusses the quantification of melanin by measuring grayscale levels, which is an effective means of assessing melanin content in fungal cultures. In this method, fungal cultures grown in the presence of L-DOPA are imaged and the grayscale

R. Pasrija, D. Kumari, *Protocols for Studying Pathogenic Fungi*,
https://doi.org/10.1007/978-981-96-7047-5_15

intensities of the images captured under the microscope are analyzed, providing a simple yet reliable approach to quantify melanin production. By understanding the mechanisms of melanin production and its quantification, this chapter aims to provide valuable insights into how melanin contributes to *Cryptococcus* pathogenicity and offers potential therapeutic targets to combat *Cryptococcus* infections.

Keywords

Cryptococcus · Immune defenses · Laccase · L-DOPA · Grayscale

Introduction This chapter contains detailed protocols ranging from melanin formation to its quantification. The topics covered include:

15.1. Materials Required
15.2. Preparations of Media and Buffers
15.3. Protocol for Melanin Formation
15.4. Protocol for Melanin Quantification
15.5. Tips and Troubleshooting

15.1 Materials Required

15.1.1 Fungal Strains and Culture Media

- **Fungal strains:** Melanin-producing *Cryptococcus* sp., *e.g. Cryptococcus neoformans* or *Cryptococcus gattii*.
- **Melanin-inducing media**
  - **L-DOPA medium:** L-3,4-dihydroxyphenylalanine (L-DOPA) is a precursor for melanin synthesis by the enzyme laccase. Dissolve it in phosphate-buffered saline (PBS) or minimal media.
  - **Niger seed agar (NSA) (optional):** Contains catecholamine precursors for melanin production and is used as a standard medium for melanin induction.
  - **Minimal media (MM):** Supplements such as asparagine or glucose can be added.

15.1.2 Laboratory Consumables

- **Petri dishes or agar plates:** Sterile plates for cultivation on solid melanin-inducing media.
- **Sterile tubes and flasks:** For liquid cultures in melanin-inducing media.
- **Pipettes and sterile tips:** For precise handling of solutions and inoculum.

15.1.3 Microscopy and Imaging Techniques

- **Light microscope**: For visualization of melanin granules in fungal cells or on plates.

15.1.4 Environmental Conditions

- **Incubation conditions**: Standard incubation at 30–37 °C.
- **Note**: 30 °C incubation increases melanin synthesis in *Cryptococcus* sp.

15.1.5 Controls

- **Positive control**: Known melanin-producing *Cryptococcus* strain.
- **Negative control**: Laccase-deficient mutant or wild-type strain grown in non-melanin-inducing media.

15.1.6 Safety and Aseptic Materials

- **Personal protective equipment**: Gloves, lab coat and goggles for safe handling.
- **Biosafety cabinet**: For working with pathogenic fungi in a controlled environment.

15.2 Preparations of Media and Buffers

15.2.1 Preparation of YEPD Media (See Chap. 2)

15.2.2 Preparation of L-DOPA Stock Solution

- **Materials required:** L-DOPA, distilled water, pH meter, autoclaved 50 mL tubes, NaOH, HCl, 0.22 μm sterile filter, *etc.*
- **Procedure for making 100 mM stock solution**
  - **Weigh L-DOPA:** Weigh 19.7 mg of L-DOPA using an analytical balance. This amount gives a 100 mM concentration when dissolved in 10 mL of distilled water.
  - **Prepare a stock solution:** Add the 19.7 mg of L-DOPA to a sterile 50 mL tube. Then, add 10 mL of distilled water to the tube and ensure the L-DOPA is completely dissolved. Ensure no L-DOPA powder remains undissolved.
  - **Sterile filtration:** Filter the solution through a 0.22 μm sterile filter to remove any particles or potential contaminants. Use sterile techniques to avoid contamination during the filtration process. Collect the filtered solution in a clean, sterile container.

- **Check the pH:** Measure the pH of the solution with a pH meter. The pH of the L-DOPA solution should be approximately 5–6, which is ideal for stability. If necessary, adjust the pH with NaOH (if too acidic) or HCl (if too basic), but avoid drastic changes that could degrade L-DOPA.
- **Storage:** Pour the filtered L-DOPA stock solution into a sterile, amber colored bottle or opaque container to protect the solution from light, which can cause L-DOPA degradation. Store the stock solution at 4 °C (refrigerator). L-DOPA is sensitive to light, so it is important to store it in a dark, cool environment to maintain its stability. The stock solution should be used within 2–3 weeks for optimal results.
- **Notes:**
  - Always prepare fresh working solutions from the stock solution when adding L-DOPA to *Cryptococcus* cultures for melanin induction.
  - To induce melanin production, dilute the 100 mM L-DOPA stock solution in growth media (*e.g.* MM or SDB) to the desired final concentration, usually 1 mM, depending on the experimental setup.
  - L-DOPA is susceptible to oxidation. Therefore, avoid exposing it to light or air for long periods of time and store it in tightly closed containers.

15.2.3 Preparation of Minimal Media (MM) [1]

- **Ingredients:** Dextrose, magnesium sulfate ($MgSO_4$), potassium hydrogen phosphate (KH_2PO_4), glycine, thiamine hydrochloride, distilled water, pH meter, autoclaved beakers, NaOH, HCl, 0.22 μm sterile filter, *etc.*
- **Protocol**
  - **Weigh the ingredients:** Weigh the following reagents to prepare 1 L of minimal media (MM): dextrose (5 mM): 9 g, $MgSO_4$ (10 mM): 2.46 g, KH_2PO_4 (30 mM): 4.2 g, glycine (13 mM): 1.0 g and thiamine hydrochloride (2.8 μM): 0.7 mg.
    Note: Weights are based on the desired final concentrations of each ingredient.
  - **Dissolve the ingredients:** Place the measured amounts of each ingredient into a clean 1 L beaker. Add distilled water to the beaker to a final volume of 1 L. Stir the solution gently with a magnetic stirrer or by hand until all solids are completely dissolved.
  - **Adjust the pH value of the medium:** Use a pH meter to check the pH of the prepared medium. The target pH value should be 5.5. Stir well after adjusting the pH value to ensure an even distribution of the acid/base.
  - **Sterilize the medium:** After preparing the media and adjusting the pH, filter the solution through a 0.22 μm sterile filter to remove microbial contaminants and particles.
  - **Storage:** Pour the filtered MM into a sterile storage container (*e.g.* an amber bottle or opaque container to protect from light) and store the medium at room temperature (25–27 °C) in a dark place to prevent degradation of sensitive components such as thiamine.

15.2.4 Preparation of PBS (See Chap. 3)

15.3 Protocol for Melanin Formation

15.3.1 Materials Required

- MM, fungal culture, PBS, centrifuge, sterile tubes, 0.85% saline solution, 12-well plate, laminar hood, *etc.*

15.3.2 Procedure

- **Inoculate the culture:** Inoculate *C. neoformans* from a fresh colony in YEPD medium. Incubate the culture for 24 h at 30 °C in a shaker set at 150 rpm.
- **Harvest the yeast cells:** Centrifuge the culture after 24 h of incubation at 3000×*g* for 5–10 min to pellet the yeast cells. Carefully decant and discard.
- **Wash the cells with PBS:** Wash cells with PBS and centrifuge the cells again at 3000×*g* for 5 min to pellet the cells. Discard the supernatant and repeat the washing step twice to ensure that the cells are thoroughly cleaned.
- **Resuspend the cells in PBS:** After the final wash step, resuspend the cells in PBS/0.85% saline to achieve an OD_{600} of 0.4 (this ensures adequate cell density for melanin induction).
- **Prepare 12-well plate:** Take a 12-well plate or culture tubes and add 2 mL of MM to each well/tube.
- **Prepare the cell suspension:** Add 50 μL of the cell suspension (OD_{600} = 0.4) to each well containing the MM.
- **Add L-DOPA:** Add 1 mM of the sterilized L-DOPA from the 100 mM stock solution to each well. Mix gently to ensure an even distribution of cells and L-DOPA in the medium.
- **Cover the plate:** Cover the plate/tube to protect it from exposure to light, as L-DOPA is photosensitive and may degrade when exposed to light. Perform the experiment in a laminar flow hood with the lights off to ensure minimal light exposure.
- **Incubate the culture:** Place the plate/tube in an incubator set to 30 °C for 3–7 days. Melanin production will be visible from the second day of incubation at 30 °C. The cultures should develop a dark color as melanin accumulates. Adjust the incubation time to your specific experimental needs. For rapid melanin formation, 3–4 days may be sufficient; for more pronounced melanin production, longer incubation times (up to 7 days) may be used.
- **Protocol for solid media (optional):** If you are using solid MM for melanin induction, follow these steps:
  - **Prepare solid media:** After adding 2 mL of MM to each well (as in step 15.3.5), add 1 mL of autoclaved agar (1.5%) to the mixture.

- **Solidify the medium:** Allow the agar to solidify in the wells.
- **Inoculate the solid media:** Add 50 μL of the cell suspension (OD_{600} = 0.4) and 1 mM L-DOPA to the solidified MM-agar mixture.
- Mix gently and incubate at 30 °C for 3–7 days, as described above.
- **NOTE:** Protect L-DOPA from light, as it is photosensitive. Carry out the experiment under low-light conditions, e.g. in a laminar flow hood with the lights switched off. Do not add L-DOPA to the negative control.

15.4 Protocol for Melanin Quantification

15.4.1 Quantification of Melanin with Adobe Photoshop CS3 and Microscopy

- **Materials required**
  - Well plates with fungal cultures grown in MM with L-DOPA for melanin induction
  - Digital camera with good image quality (preferably with adjustable light settings)
  - Adobe Photoshop CS3 (or similar software with grayscale analysis tools)
  - Light microscope (100× magnification)
  - Microscope slides and cover glasses
  - Pipettes and sterile tips for transferring samples
  - PBS for washing the cells (optional for preparing the samples for microscopy)
- **Acquire digital images for grayscale analysis**
  - Take images of the wells in which *C. neoformans* was cultured with L-DOPA under uniform lighting conditions.
  - Ensure that the camera settings are consistent for all samples, including white balance, focus, and light intensity, to avoid variations in image quality.
  - Place the wells or tubes in the same lighting setup to ensure even exposure and avoid over or underexposure. This ensures accurate grayscale measurement.
  - Take high quality photos of the wells. Ensure that the entire surface of the well or section is in focus.
  - For consistent results, always maintain the same distance between the camera and the wells to maintain scale and resolution.
- **Quantification of melanin with Adobe Photoshop CS3**
  - **Open the image in Photoshop:** Open the captured image of the culture in Adobe Photoshop CS3.
  - **Convert the image to grayscale:** Convert the image to grayscale by selecting Image > Mode > Grayscale. If your image contains color, this step will ensure that only the intensity of the gray tones representing the pigmentation is considered.
  - **Measure the intensity of pigmentation**

- In Photoshop, select the eyedropper tool (shortcut: I) to click on the region of interest (i.e. the area of the well where melanin pigmentation is present).
- Once you have selected a pixel in the region of interest, the Color Picker window will display the gray value (brightness) of that pixel in the "Gray" channel. The value ranges from 0 (black) to 255 (white).
- Average Gray Value: Use the eyedropper tool to sample several areas of the image, especially in regions with visible pigmentation, and record the average gray value.
- You may need to use the Histogram tool (Window > Histogram) to get a more accurate impression of the overall distribution of gray levels in the image.

– **Calculate the percentage of pigmentation:** Use the following formula to calculate the percentage of pigmentation:

$$\text{Pigmentation}\% = \left[\left(255\,\text{subtracted from the mean value of the test image}\right)/\,255\right]\times 100$$

– The mean gray value should be calculated from the test sample (the region with melanin), and 255 represents the lightest (whitest) possible color in shades of gray. The pigmentation % gives you an estimate of the melanin content in the sample based on how dark the regions of the culture appear in the image. A higher percentage indicates greater melanin production (darker pigmentation).

– **Document and compare the results:** Record the percentage of pigmentation for each sample and compare it between experimental groups (*e.g.* treated vs. control). You can also compare different time points to track changes in melanin production.

15.4.2 Microscopic Observation for Melanin Localization

- **Prepare the slide for microscopy**
  - Remove 5 µL of melanized cells or control cells from the culture.
  - Place the sample on a clean microscope slide.
  - If necessary, wash the cells with PBS to remove excess medium before placing them on the slide.
  - Place a coverglass/coverslip over the sample to protect it and to better focus the cells under the microscope.
- **Microscopic observation**
  - Examine the prepared slide under a light microscope using 100× magnification (oil immersion objective).
  - Look for melanin as dark pigmentation around the cells, appearing as black or brown areas within the fungal cells (as shown in Fig. 15.1).
- **Documentation**
  - Take representative images of the melanized cells under the microscope using a digital camera attached to the microscope.
  - Record the images for later reference or comparative analysis.

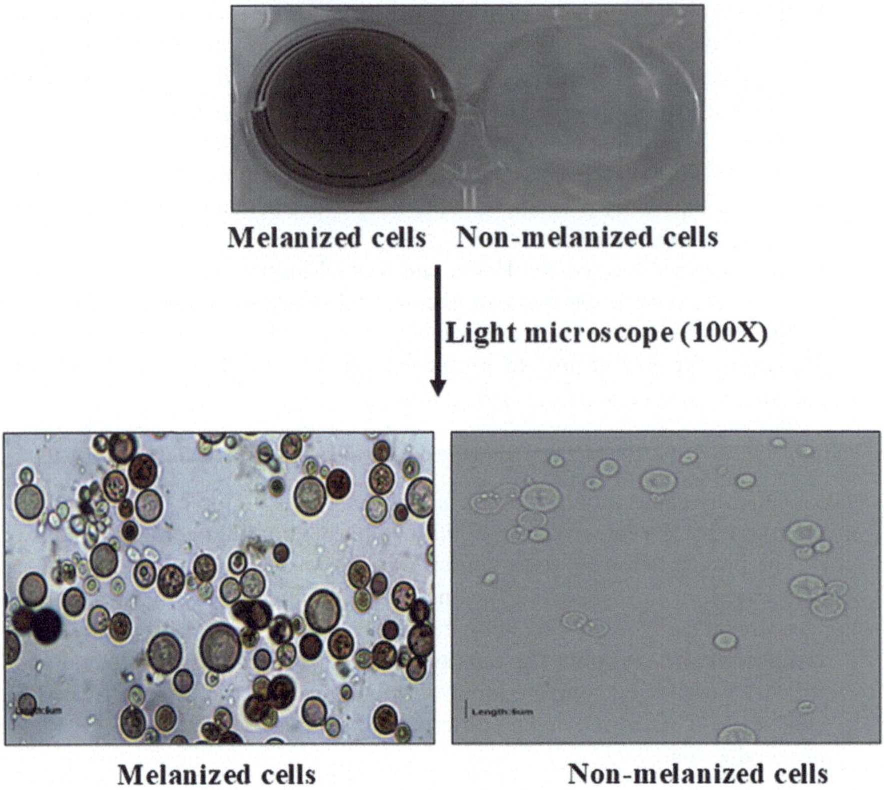

Fig. 15.1 Images of melanized and non-melanized wells and light microscopic views at 100× magnification

15.5 Tips and Troubleshooting

15.5.1 Tips

1. **Selection of the fungal strain**
   - Use melanin-producing strains such as *Cryptococcus neoformans* H99.
   - Ensure that the strains are viable and contamination-free before starting the test.
2. **Media and substrate for melanin induction**
   - Melanin production requires substrates such as L-DOPA, which serve as precursors for melanin synthesis.
   - **Recommended media**
     - **L-DOPA agar:** Prepare media with 1 mM L-DOPA in a minimal medium.
     - **Niger seed agar (NSA):** which naturally contains catecholamines
     - **Control media:** Add a medium without L-DOPA or substrate as a negative control to confirm substrate-dependent melanin production.

3. **Preparation of the inoculum**
   - Use cells in the mid-logarithmic phase for optimal melanin production.
   - Wash cells in sterile PBS to remove residual nutrients or media components that may interfere with induction.
4. **Incubation conditions**
   - **Temperature:** Incubate the cultures at 30 °C.
   - **Time:** Allow melanin production for 3–7 days, depending on the medium and strain.
   - **Dark environment:** It is imperative to keep the plates in the dark to minimize photo-oxidation of L-DOPA.

15.5.2 Troubleshooting

1. **Problem: No melanin formation (no dark color change)**
   - **Possible causes:**
     - The fungal cells may not be metabolizing L-DOPA efficiently.
     - The cells may not be viable, or the L-DOPA concentration may be too low.
     - Unsuitable incubation conditions (temperature, exposure to light, or time) could affect melanin formation.
     - The pH may not be optimal for melanin production.
   - **Solutions:**
     - Ensure that the *Cryptococcus* strain is capable of melanin production. Some strains may lack this ability.
     - Check the quality and viability of the fungal culture by ensuring proper growth and colony formation in YEPD before testing.
     - Ensure the L-DOPA is freshly prepared and at the correct concentration (1 mM).
     - Incubate the cultures for 3–7 days at 30 °C and ensure that the cultures are covered and protected from light to prevent degradation of L-DOPA.
     - If using solid medium, ensure the agar is evenly mixed and well solidified.
     - Check and adjust the pH of the medium and adjust it to ensure it is in the correct range (approximately pH 5.4–5.7). An alkaline pH can lead to premature polymerization of L-DOPA, and a high acidic pH could inhibit melanin formation.
2. **Problem: Dark color change appears too early**
   - **Possible causes:**
     - Self-polymerization of L-DOPA could occur, especially at alkaline pH, resulting in premature darkening of the medium.
   - **Solutions:**
     - Make sure that the pH of the MM medium is in the slightly acidic range (around pH 5.4–5.7). An alkaline pH value promotes the non-specific self-polymerization of L-DOPA.
     - To prevent the self-polymerization of L-DOPA, add ascorbic acid (5 mM final concentration) to the medium. Ascorbic acid acts as an antioxidant and helps to stabilize L-DOPA and prevent its non-specific polymerization.

   - Store L-DOPA and the medium under low-light conditions to avoid light-induced degradation.

3. **Problem: Uneven melanin formation (inconsistent dark zones)**
   - **Possible causes:**
     - Uneven inoculation of fungal cells or improper mixing of L-DOPA and cells could lead to inconsistent melanin formation.
     - The agar may not have solidified properly, resulting in an uneven surface distribution.
   - **Solutions:**
     - Inoculate the plates with a uniform volume of fungal suspension and ensure the cell suspension is evenly mixed. Use a sterile pipette or inoculation loop to distribute the cells evenly.
     - Make sure that the agar solidifies evenly and that there are no air bubbles or inconsistencies in the media.
     - If using liquid media, ensure the cell suspension and L-DOPA solution are well mixed before incubation.
     - Incubate the plates in a stable environment at 30 °C with minimal agitation to avoid uneven growth.
4. **Problem: Melanin formation in the controls (negative control shows dark color)**
   - **Possible causes:**
     - The negative control (fungi without laccase activity or without L-DOPA) may show unexpected dark coloration due to contamination, excessive exposure to light or improper preparation of the medium.
   - **Solutions:**
     - Ensure that the negative control medium does not contain L-DOPA and is properly sterilized. Also, check the pH of the control plates to ensure that they remain within the desired range and do not induce non-specific polymerization.
     - Always store the plates in the dark and protect them from excessive exposure to light to avoid premature polymerization of L-DOPA in the control group.
     - Check the plates for contamination with other fungal species, as other organisms may have laccase activity or produce pigments that could mimic melanin formation.
5. **Problem: Poor reproducibility of results (different results between replicates)**
   - **Possible causes:**
     - Different incubation conditions (temperature, humidity, light) or inconsistent handling of samples can lead to differences in melanin production.
     - Inconsistent cell suspension concentrations can lead to unequal growth rates in replicates.
   - **Solutions:**
     - Ensure all wells or plates are incubated under identical conditions, with stable temperature and minimal light exposure.

- Prepare all cultures at the same cell density (OD_{600} = 0.4) and resuspend cells thoroughly before inoculation.
- If using liquid media, ensure that the medium is evenly mixed with the L-DOPA solution and avoid variations in volume or pH between replicates.
- Record and standardize any variations in experimental setup or environmental conditions during the experiment.

6. **Problem: Inconsistent representation of melanin formation**
   - **Possible causes:**
     - Varying lighting conditions during image acquisition can lead to inconsistent or inaccurate melanin quantification.
     - Inappropriate camera settings (*e.g.* exposure, focus, white balance) can distort image quality.
   - **Solutions:**
     - Take all images under consistent lighting conditions and with the same camera settings (white balance, focus, exposure) for each replicate.
     - Always use a high-quality camera with good resolution to capture clear images, and maintain the same distance from the wells to avoid scale variations.
     - For accurate quantification using Adobe Photoshop or other software, ensure that images are captured with sufficient resolution and clarity to detect subtle variations in melanin production.
7. **Problem: No color change despite expected melanin production**
   - **Possible causes:**
     - L-DOPA of inferior quality or improper storage conditions can lead to a lack of melanin production.
     - The fungal strain may have defective melanin biosynthesis pathways or an inadequate ability to process L-DOPA.
     - The incubation temperature or pH may not be optimal for melanin production.
   - **Solutions:**
     - Use a freshly prepared L-DOPA stock solution and ensure it is stored properly (avoid light and extreme temperatures).
     - Ensure that the fungal strain is melanin-competent (some *Cryptococcus* strains may lack the ability to produce melanin). Confirm this with a known melanin-producing strain (*e.g. C. neoformans* wild type) for comparison.
     - Check that the incubation temperature (30 °C) is optimal for melanin production. If necessary, adjust the incubation time (3–7 days) to achieve greater melanin accumulation.
     - Note that for *C. neoformans* H99, a higher pH and incubation at 30 °C promote melanin production [2]. However, this may not be true for all fungal species. The optimal conditions for melanin production (temperature and pH) may vary from fungus species to fungus species. Therefore, always check and optimize the conditions for the fungal species studied.

References

1. Brilhante RSN, España JDA, de Alencar LP et al (2017) An alternative method for the analysis of melanin production in *Cryptococcus neoformans sensu lato* and *Cryptococcus gattii sensu lato*. Mycoses 60:697–702. https://doi.org/10.1111/myc.12650
2. Kumari D, Sachivkina N, Pasrija R (2024) Investigation of the influence of pH and temperature on melanization and survival under oxidative stress in *Cryptococcus neoformans*. Arch Microbiol 206(8):355. https://doi.org/10.1007/s00203-024-04080-7

Urease Assay for *Cryptococcus* sp.

16

Abstract

Urease is an extracellular enzyme in various bacteria fungi, and plants that plays a crucial role in the pathogenicity of various microorganisms. *Helicobacter pylori,* for example, uses urease to hydrolyze urea. This produces ammonia, which increases the pH of its environment and allows the bacterium to survive in the acidic gastric mucosa, contributing to the development of gastric ulcers (Graham, Miftahussurur, J Adv Res 13:51–57, 2018). Many human pathogenic fungi, including *Cryptococcus neoformans*, *Candida albicans*, *Coccidioides immitis*, *Histoplasma capsulatum*, *Sporothrixschenckii* and species of *Trichosporon* also exhibit urease activity (Feder et al, FEBS J 282:1406–1418, 2015; Cox et al, Infect Immun 68:443–448, 2000). In *C. neoformans*, urease production during infection leads to the formation of ammonia, which alkalizes the tissue environment and contributes to tissue damage. This enzymatic activity has been linked to the ability of *C. neoformans* to cross the blood-brain barrier in mouse models of cryptococcosis and promotes invasion of the central nervous system (Toplis et al, FEMS Yeast Res 20:foaa031, 2020). Urease is a metalloenzyme that catalyzes the hydrolysis of urea to ammonia and carbon dioxide, increasing the pH of the surrounding medium. The pH shift can be visually detected by a color change of the phenol red indicator from light orange (pH 6.8) to pink (pH 8.1). Fungi with high urease activity usually cause the medium to turn pink within 24 h, and the intensity of the pink coloration correlates directly with the degree of urease activity (Cox et al, Infect Immun 68:443–448, 2000). This chapter describes a detailed protocol for measuring urease activity in pathogenic fungi, using urea agar plates as the medium for this assay. In this method, the appearance of a pink zone around the fungal colonies on the urea agar serves as an indicator of urease activity. The size and intensity of the pink zone can be quantitatively assessed to measure the level of urease production. This chapter describes a protocol for measuring urease activity in fungi.

R. Pasrija, D. Kumari, *Protocols for Studying Pathogenic Fungi*,
https://doi.org/10.1007/978-981-96-7047-5_16

Keywords

Urease · Fungi · Murine model · Ammonia · *Candida albicans*

Introduction This chapter deals with detailed protocols for measuring urease activity in fungi. Topics covered include:

16.1. Materials Required
16.2. Preparations of Media and Buffers
16.3. Protocol
16.4. Tips and Troubleshooting

16.1 Materials Required

16.1.1 Fungal Strains and Growth Media

- **Fungal strains:** Urease-positive *Cryptococcus* sp. (*e.g. Cryptococcus neoformans* or *Cryptococcus gattii*) and urease-negative strain (optional, as negative control).
- **Culture media:** Yeast extract peptone dextrose (YEPD)
- **Urea agar base or urea broth:** Prepared with urea as a substrate. Commercially available urea agar base (*e.g.* Christensen's urea agar) is commonly used and contains a pH indicator.

16.1.2 Reagents for Detection of Urease Activity

- **Urea:** Analytical-grade urea, for the preparation of urea-containing media or solutions.
- **pH indicator (Phenol red):** Usually contained in urea agar/broth; changes color due to pH shift due to ammonia production.
  - Acidic pH: Yellow.
  - Alkaline pH (urease activity): Pink to red.
- **Buffer solution:** Phosphate buffered-saline (PBS) (for washing fungal cells).

16.1.3 Laboratory Consumables

- **Petri dishes or test tubes**: To prepare solid or liquid media.
- **Sterile pipettes and tips**: For precise handling of liquid media.
- **Sterile tubes or flasks**: For the preparation of fungal suspensions.

16.1.4 Instruments

- **Spectrophotometer (optional)**: For determining the required optical density (OD_{600}) of the culture
- **Incubator**: To maintain optimal growth conditions (30 °C for *Cryptococcus* sp.).

16.1.5 Controls

- **Positive control**: A known urease-positive strain of *Cryptococcus* sp.
- **Negative control**: Urease-negative mutant or medium without urea.

16.1.6 Environmental and Safety Materials

- **Personal protective equipment**: Gloves, lab coat, and safety goggles.
- **Biosafety cabinet:** For the safe handling of pathogenic fungi.

16.2 Preparations of Media and Buffers

16.2.1 Preparation of the YEPD Media (See Chap. 2)

16.2.2 Preparation of 0.85% Saline Solution (See Chap. 3)

16.2.3 Preparation of PBS (See Chap. 3)

16.2.4 Preparation of Christensen's Urea Agar Plates

- **Materials required:** Peptone (1 g), dextrose (1 g), NaCl (sodium chloride) (5 g), disodium phosphate (Na_2HPO_4) (1.2 g), monopotassium phosphate (KH_2PO_4) (0.8 g), phenol red (0.012 g), agar (15 g), Urea 10 g (for the preparation of the stock solution of urea solution), distilled water, autoclave, sterile filter (0.22 μm) (for urea sterilization) and sterile Petri dishes.
- **Procedure**
  - **Prepare the urease agar base** [1]: In a large beaker or flask, mix the following components: 1 g peptone, 1 g dextrose, 5 g NaCl, 1.2 g Na_2HPO_4, 0.8 g KH_2PO_4, 0.012 g phenol red, 15 g agar. Add distilled water (to make the volume 1 L) to the beaker/flask containing the above ingredients. Heat the mixture to boiling while stirring to completely dissolve all components. Make sure that the agar is completely dissolved and the solution is homogeneous.
  - **Sterilization of the urease agar base:** Autoclave the solution at 121 °C for 15 min to sterilize the agar medium and remove possible impurities.

- **Preparation of the urea solution:** Dissolve 10 g of urea in distilled water and adjust the final volume to 100 mL to prepare a 10% urea solution. Filter to sterilize the urea solution with a 0.22 μm sterile filter to remove microorganisms and maintain sterility.
- **Combine the urea solution with the urease agar base:** Allow the agar base to cool to about 50–60 °C after autoclaving. The agar should still be liquid but cold enough to add the urea solution without degrading the components. Add 2% sterile urea solution (from the 10% stock) to the cooled agar base, mixing thoroughly. Check the pH of the final mixture. The pH should be approximately 6.8 ± 0.1. If necessary, adjust the pH with diluted HCl (to lower the pH) or diluted NaOH (to raise the pH).
- **Pour agar plates:** Pour the prepared agar aseptically sterile Petri dishes, about 20–25 mL per plate, depending on the size of the dish. Allow the agar plates to cool and solidify at room temperature.
- **Storage and inoculation:** Store the solidified agar plates at 4 °C temperature in a dry, sterile environment, protected from light, until ready for use. Once ready, you can inoculate the plates with the desired fungal species to test for urease activity.
- **Note**
  - **Sterility**: Ensure that all materials and tools used for preparation, especially when adding the urea solution, are sterile to avoid contamination.
  - **Phenol red sensitivity**: Phenol red serves as a pH indicator. It turns from yellow at pH 6.8 to pink at pH 8.1 when urease activity is present (alkalinization by ammonia production).
  - **Urease activity**: Urease positive organisms color the medium around the colony pink after 24–48 h of incubation at the appropriate temperature (typically 30 °C for fungi).

16.3 Protocol

16.3.1 Materials Required

- YEPD media, PBS, saline, Christensen's urea agar plates, fungal culture and incubator, ruler, camera, *etc.*

16.3.2 Procedure

- **Preparation and growth of yeast cells**
  - Inoculate yeast cells (*e.g. C. neoformans*) in YEPD medium and allow them to grow overnight at 30 °C in a rotary shaker at 150 rpm for 24 h.
  - Centrifuge the yeast culture after 24 h at 3000×*g* for 5 min to collect the cells.
  - Remove the supernatant and wash the cell pellet twice with PBS or 0.85% saline to remove any residual medium or contaminants.

  - Resuspend the washed yeast pellet in 0.85% saline at a concentration of 10^6 cells/mL (determined by optical density or direct cell counting with a hemocytometer or spectrophotometer).
- **Inoculation of Christensen's urea agar plates**
  - Ensure that Christensen's urea agar plates are yellow-orange in color before inoculation. These plates should be stored at 4–8 °C until needed. However, do not reheat them as the urea will decompose.
  - Inoculate 10 μL of the yeast suspension (10^6 cells/mL) into the center of each Christensen's urea agar plate using a sterile pipette tip or a 10 μL inoculation loop.
  - **Note:** Ensure the inoculum is applied directly to the agar surface and that the volume of inoculum is evenly distributed. Avoid excessive spreading to prevent merging with other colonies.
  - Place the inoculated plates in an incubator set at 30 °C and incubate for 2–3 days. The incubation time may vary depending on the yeast strain; check regularly for color changes.
- **Observation and measurement of urease activity**
  - Examine the plates after 2–3 days for a change in color of the medium surrounding the inoculum. A pink zone around the yeast colonies indicates urease activity (as shown in Fig. 16.1), where ammonia production from urea hydrolysis has increased the pH of the medium.
  - **Positive urease activity:** The medium around the colonies turns pink (indicating a pH shift to ~8.1).
  - **Negative urease activity:** If there is no color change or the medium remains yellow-orange, urease activity is not present.
  - Measure the diameter of the pink zone around the colony with a ruler. This zone reflects the extent of urease activity; a larger zone indicates higher urease production.

16.4 Tips and Troubleshooting

16.4.1 Tips

1. **Selection of the fungal strain**
   - Use urease-positive strains, such as *Cryptococcus neoformans* H99.
   - Include urease-deficient mutants or strains as negative controls to confirm the assay's specificity.
2. **Preparation of the medium**
   - **Christensen's urea agar/broth**
     - Prepare a medium containing urea (*e.g.* Christensen's urea agar or broth with phenol red as pH indicator).
     - Ensure the urea concentration is within 1–2% (w/v) for optimal detection.
     - Use a sterile, fresh medium, as urea can degrade over time.

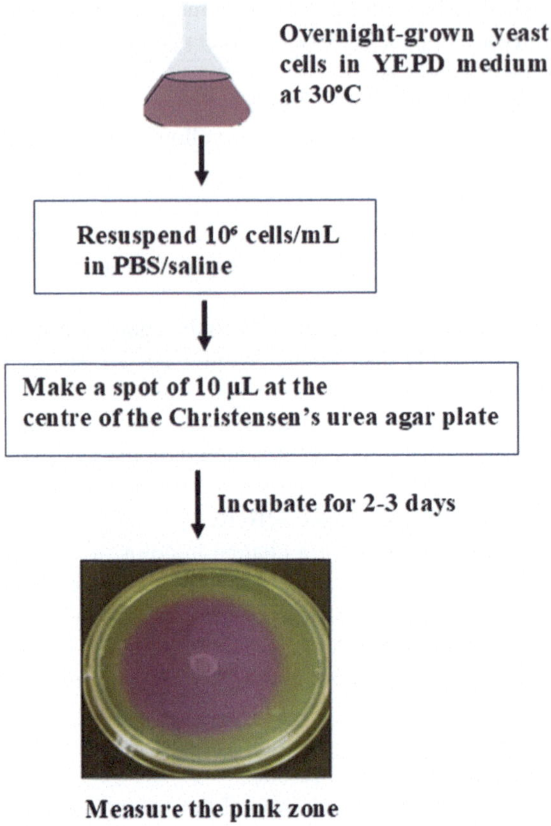

Fig. 16.1 The urease activity of *C. neoformans* was determined by spotting the cells on Christensen's agar plates and incubating them at 30 °C for 48 h. Urease activity was determined by measuring the diameter of the pink halo around the inoculum, which results from the incorporation of the pH indicator phenol red

- **pH value of the medium**
  - Adjust the pH of the medium to approximately 6.8 ± 0.1 to allow clear detection of pH changes.

3. **Preparation of the inoculum**
   - Use cells in the mid-logarithmic phase grown in YEPD or minimal medium.
   - Wash the cells with sterile PBS to remove residual nutrients.
   - Standardize the inoculum density (*e.g.* 10^6–10^7 cells/mL) for uniform results.
4. **Incubation conditions**
   - **Temperature:** Incubate at 30 °C depending on the experimental setup.
   - **Time:** Allow urease activity for 48–72 h; observe color changes at regular intervals.
   - **Aeration:** For liquid assays, shake gently to maintain aeration and ensure adequate enzyme activity.

16.4.2 Troubleshooting

1. **Problem: No color change (no pink zone)**
   - **Possible causes**:
     - The fungal strain may not be urease positive.
     - Insufficient inoculum or low fungal concentration.
     - Insufficient incubation time or temperature.
     - The urea in the agar may have degraded due to improper storage.
   - **Solutions:**
     - **Confirm a urease-positive strain**: Ensure that the fungal species tested is urease positive. Use a positive control (*e.g. Proteus vulgaris, C. neoformans*) to check the sensitivity of the assay.
     - **Enlarge the inoculum**: Make sure that the inoculum is sufficient and properly resuspended. For a visible urease test, aim for 10^6 cells/mL.
     - **Incubate for longer or at a constant temperature**: Incubate plates at 30 °C and check after 48–72 h. Some strains may require a longer incubation time to show urease activity.
     - **Check the urea solution**: Make sure the urea solution is freshly prepared and properly sterilized. If stored improperly or for too long, the urea may degrade and reduce its availability for hydrolysis.
     - **Check the phenol red sensitivity**: The phenol red indicator should be freshly mixed into the agar. If the phenol red is old or degraded, it may not show the expected color change.
2. **Problem: False positive: Color change around non-urease-producing fungi**
   - **Possible causes**:
     - Contamination with urease-positive organisms during preparation or inoculation.
     - Incorrect pH value of the agar medium before inoculation.
     - Decomposition of the urea leads to a natural increase in pH of the medium.
   - **Solutions:**
     - **Check sterility**: Ensure media, pipettes and inoculation loops are sterile to prevent contamination with urease-positive organisms. Always take negative controls (non-urease-producing fungi) to check the results.
     - **Pay attention to the correct pH**: To avoid false results, it is important to know that an alkaline pH shift in the agar can cause a color change from yellow to pink, which is not necessarily due to urease activity. This color change can occur when the pH of the medium becomes alkaline for reasons unrelated to the production of ammonia by urease enzymes. The pH of the medium should be about 6.8 ± 0.1 before adding the urea solution. If the pH is incorrect, adjust it with diluted HCl or NaOH before autoclaving.
     - **Inoculation technique**: Avoid excessive inoculation, which could spread the decomposition of the urea or allow cross-contamination between colonies. Use a controlled inoculum size.

3. **Problem: Weak or inconsistent pink zone**
   - **Possible causes**:
     - Inconsistent inoculation or uneven distribution of fungal cells.
     - Inadequate incubation conditions (temperature fluctuations or insufficient time).
     - Excessively thick agar or poorly mixed media.
   - **Solutions**:
     - **Standardize the size of the inoculum**: Ensure that the fungal inoculum is applied uniformly at 10^6 cells/mL. Using a sterile inoculation loop or pipette can help with uniform distribution.
     - **Control the incubation conditions**: Ensure a constant incubation at 30 °C. Check that the temperature of the incubator is stable, as fluctuations can affect urease activity.
     - **Check the media preparation**: Ensure the medium is well-mixed and lumps-free. The agar should be poured evenly into the plates, and the medium should not be too thick as this may interfere with the detection of urease activity.
4. **Problem: The plates are not yellow-orange before inoculation**
   - **Possible causes:**
     - Incorrect preparation or incorrect pH of the medium before adding the urea solution.
     - Decomposition of the urea due to improper storage or handling of the plates.
   - **Solutions:**
     - **Check the pH value**: Ensure the pH value of the agar is around 6.8 (yellow-orange) before inoculation. If necessary, adjust the pH before adding the urea solution.
     - **Check for correct storage**: Store the prepared plates at 4–8 °C and protect them from light. Do not reheat the prepared plates as the urea may decompose and affect the test result.
5. **Problem: Agar plates show overgrowth or fusion of colonies**
   - **Possible causes:**
     - Too much inoculum or colonies too close together.
     - Overgrowth of the fungal culture due to prolonged incubation or high cell density.
   - **Solutions:**
     - **Reduce the size of the inoculum**: Add5–10 μL of the 10^6 cells/mL yeast suspension to prevent excessive growth of the colonies. Avoid over-inoculation, as this could lead to fusion of the colonies.
     - **Distribute the colonies**: When inoculating multiple plates, make sure the colonies are spaced far enough apart to prevent merging. This also allows better visualization of the individual urease activity zones.
6. **Problem: Contamination or inaccurate results due to poor handling**
   - **Possible causes**:
     - Non-sterile conditions during media preparation or inoculation.
     - Cross-contamination between fungal species.

- **Solutions:**
  - **Ensure sterile technique**: Always use sterile pipettes, inoculation loops and clean surfaces to avoid contamination.
  - **Use appropriate controls**: Include positive (*e.g. Proteus vulgaris* or *C. neoformans* H99) and negative (*e.g.* non-urease-producing fungus) controls to confirm the accuracy of the test.

References

1. Gutierrez-Gongora D, Raouf-Alkadhimi F, Prosser RS, Geddes-McAlister J (2023) Differentiated extracts from freshwater and terrestrial mollusks inhibit virulence factor production in *Cryptococcus neoformans*. Sci Rep 13:4928. https://doi.org/10.1038/s41598-023-32140-3

Phosphlipase Assay for the Detection of Virulence of Pathogenic Fungi

17

Abstract

Microbial pathogens, including bacteria and fungi, utilize various genetic strategies to invade host tissues and cause infections. One of these strategies is the secretion of phospholipases, a class of enzymes that degrade phospholipid components of host cell membranes, destroying their structural integrity and facilitating tissue invasion (Ghannoum, Clin Microbiol Rev 13:122–143, 2000). It is known that pathogenic fungi such as *Candida* sp./*Cryptococcus* sp. secrete phospholipases that play a crucial role in their virulence and ability to evade host immune responses. These enzymes hydrolyze phospholipids, leading to the formation of bioactive metabolites that can damage host cells and tissues. Phospholipase activity is particularly associated with fungal pathogenesis. It contributes to the ability of these organisms to colonize and infect various human tissues, including mucous membranes, skin and internal organs (Djordjevic, Front Microbiol 1:125, 2010). The principle of detecting phospholipase activity in fungi is based on the hydrolysis of phospholipids in egg yolk agar. Egg yolk agar is rich in phospholipids, which, when degraded by phospholipases, form clear zones or halos around the fungal colonies (Sharaf et al, J Taibah Univ Sci 8:207–215, 2014). This clear zone indicates phospholipase activity and is measured as the phospholipase production zone (Pz). The diameter of this zone correlates with the degree of enzyme activity and provides a reliable method for assessing the pathogenic potential of fungi (Jain et al, Int J Med Sci Public Health 1, 2017). The degradation of lecithin forms a clear zone, an important phospholipid in egg yolk, which gets hydrolyzed to insoluble products that precipitate and leave a transparent area around the fungal growth. This chapter describes a detailed protocol for the determination of phospholipase activity in fungal isolates on egg yolk agar. The method described can be applied to various fungal species and helps researchers investigate phospholipases' role in fungal virulence, invasion, and infection.

R. Pasrija, D. Kumari, *Protocols for Studying Pathogenic Fungi*,
https://doi.org/10.1007/978-981-96-7047-5_17

Keywords

Phospholipase · Fungi · Phospholipids · Egg yolk · Pz · Clear zone

Introduction This chapter presents detailed protocols for measuring phospholipase activity in fungi. Topics covered include:

17.1. Materials Required
17.2. Preparations
17.3. Detailed Protocol
17.4. Tips and Troubleshooting

17.1 Materials Required

17.1.1 Fungal Strains and Growth Media

- **Fungal strains:** Pathogenic fungi known for phospholipase activity (*e.g. Cryptococcus* sp./*Candida* sp.). Strains with reduced or no phospholipase activity as controls (optional).
- **Pre-culture media:** Yeast extract peptone dextrose (YEPD)

17.1.2 Specialized Phospholipase-Inducing Media

- **Egg yolk agar**: The key medium for testing phospholipase activity assay. Prepare it as follows:
  - **Base medium**: Sabouraud dextrose agar (SDA)
  - **Egg yolk emulsion:** Commercially available sterile egg yolk emulsion or freshly prepared egg yolk. Add it to the base medium (8–10%) after autoclaving and cooling to 50 °C.
  - **Calcium chloride ($CaCl_2$):** Additive to improve the detection of phospholipase activity.

17.1.3 Reagents and Chemicals

- **Sterile Water or Phosphate-Buffered Saline (PBS):** For the preparation of fungal suspensions.

17.1.4 Laboratory Consumables

- **Petri dishes**: For pouring the egg yolk agar medium.
- **Sterile pipettes and tips**: For inoculation of fungal cells.
- **Sterile tubes or flasks**: For the preparation of fungal suspensions.
- **Calipers or ruler:** For measuring the precipitation zone.

17.1.5 Instruments

- **Incubator**: To maintain optimal growth conditions, normally 30–37 °C.

17.1.6 Controls

- **Positive control**: A strain known to produce phospholipase.
- **Negative control**: A strain without phospholipase activity or a medium without egg yolk.

17.1.7 Safety and Aseptic Materials

- **Personal protective equipment (PPE)**: Laboratory coat, gloves and safety glasses.
- **Biosafety cabinet**: For the safe handling of pathogenic fungi.
- **Autoclave or sterilizer**: For sterilizing media, tools, and waste.

17.2 Preparations

17.2.1 Preparation of YEPD Media (See Chap. 2)

17.2.2 Preparation of PBS (See Chap. 3)

17.2.3 Preparation of Egg Yolk Agar Plates

Materials Required SDA, sodium chloride (NaCl), calcium chloride ($CaCl_2$), egg yolk, distilled water, autoclave for sterilization, sterile Petri plates, centrifuge (for egg yolk separation).

- **Preparation of the egg yolk medium**
  - Prepare the SDA base for 1 L medium (dissolve 15 g SDA in about 900 mL of distilled water).

  - Mix this SDA base solution with 58.44 g NaCl (to obtain 1 M NaCl) and 0.725 g $CaCl_2$ (to obtain 5 mM $CaCl_2$) in 1 L medium.
  - After adding the NaCl and $CaCl_2$, the total volume of the solution would be slightly less than 1 L (due to the addition of the solid components).
  - Add distilled water to the mixture to bring the final volume to exactly 1 L.
  - Mix thoroughly to ensure even distribution of all components.
  - Sterilize the medium by autoclaving at 121 °C for 15 min to ensure that it is free from microbial contamination.
  - Prepare the egg yolk: Centrifuge the egg yolk at 5000×*g* for 10 min to separate the yolk supernatant (the liquid part) from the solid components at 4 °C.
- **Preparation of the egg yolk agar plates**
  - After autoclaving, allow the sterilized SDA base to cool to about 45–50 °C (lukewarm).
  - Add the appropriate amount of egg yolk supernatant to the cooled, sterilized SDA medium to achieve an 8% egg yolk concentration in the final medium. The specific volume of egg yolk added will depend on the amount of medium prepared, but for 1 L of agar, about 80 mL of egg yolk supernatant is added.
  - Stir the medium gently to ensure the egg yolk is evenly distributed in the agar base.
- **Pour the plates**
  - Pour the prepared egg yolk medium into sterile Petri plates. The volume should be about 15–20 mL per plate, depending on the size of the Petri dish.
  - Let the agar set at room temperature in a sterile environment.
- **Storing the prepared plates**
  - Once set, store the egg yolk agar plates in the refrigerator at 4 °C until needed. These plates should be used within 2–3 weeks to achieve optimal results.
  - Do not store them for extended periods of time as the egg yolk may start to break down.
- **Notes**
  - **Egg yolk supernatant:** Ensure that only the clear supernatant of the centrifuged egg yolk is used, as the solid egg yolk components can affect the consistency and clarity of the agar.
  - **$CaCl_2$:** Calcium is added to promote phospholipase activity. It is necessary for the enzymatic hydrolysis of the phospholipids contained in the egg yolk.
  - **Sterility:** Handle all components under sterile conditions to avoid media contamination and ensure accurate results.

17.3 Protocol

- **Preparation of the yeast culture**
  - **Culturing yeast:** Inoculate the YEPD medium with the fungal isolate (yeast) and incubate for 24 h in a rotary shaker at 30 °C and 150 rpm. This ensures sufficient growth of the yeast.

- **Harvest yeast cells**: Centrifuge the culture after 24 h at 3000×*g* for 5 min to pellet the yeast cells.
- **Wash the yeast cells:** Carefully remove the supernatant and resuspend the yeast pellet in PBS. Centrifuge the suspension again and discard the supernatant. Repeat the washing procedure twice to ensure that the remaining medium components are completely removed.
- **Resuspend the yeast cells:** Following the second wash, resuspend the yeast cells in PBS to obtain a final concentration of 10^6 cells/mL.

• **Inoculation of egg yolk agar plates**
  - **Inoculation**: Using a sterile pipette or loop, spot inoculate 10 μL of the yeast suspension (10^6 cells/mL) onto the surface of the egg yolk agar plates. The inoculum should be applied to the center of the plate to allow sufficient growth and observation of the precipitation zone.

• **Incubation**: Place the inoculated plates in an incubator at 30 °C/37 °C and incubate for 3–5 days. The optimal incubation time is usually 3–5 days but may vary based on the growth rate of the fungus.

• **Evaluation of the results**
  - **Observation of the precipitation zone (Pz):** Visually inspect the petri dishes after the incubation period for the formation of Pz around the fungal colony.
  - **Precipitation zone**: This is a clear or opaque zone around the colony, indicating the breakdown of phospholipids in the egg yolk by the phospholipase enzyme produced by the fungal isolate (as shown in Fig. 17.1)
  - **Measuring the zone**
    Measure the colony's diameter and the diameter of the precipitation zone formed around it. This can be done with a ruler.
  - **Calculate the Pz ratio (Pz calculation formula)**:

    Colony diameter / Colony diameter + Precipitation zone diameter

  - **Interpret Pz values:**

Pz ≥ 0.64 to <1:	High phospholipase activity
Pz < 0.64:	Very high phospholipase activity
Pz = 1:	No phospholipase activity

  - **Note:** The lower the Pz value, the greater the phospholipase production, as a larger precipitation zone corresponds to greater enzymatic activity.
  - **Triplicates for accuracy:** Perform the test in triplicate to ensure accuracy and reproducibility of results. This allows you to confirm the consistency of the precipitation zones across multiple replicates.
  - **Control samples (optional)**: It is recommended to include a positive control (a known phospholipase-producing strain) and a negative control (*e.g.* a strain known not to produce phospholipase) to verify that the assay is working as expected.

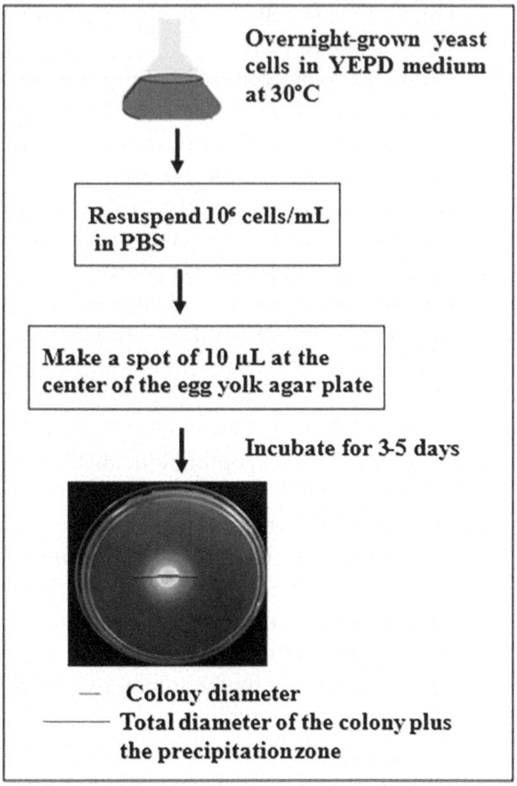

Fig. 17.1 The extracellular phospholipase activity of *C. albicans* was assessed by culturing on egg yolk agar at 37 °C for 4 days and measuring the Pz in the vicinity of the colony. Pz = ¼ = 0.25, and if the Pz value is less than 0.64, this indicates a very high phospholipase activity

17.4 Tips and Troubleshooting

17.4.1 Tips

1. **Selection of the fungal strain**
   - Use virulent fungal strains that are known to secrete phospholipases, such as *Candida* sp./*Cryptococcus* sp.
   - Ensure the strains are grown and stored under optimal conditions to avoid loss of virulence traits.
2. **Medium preparation**
   - **Phospholipid-containing media**: Use agar plates enriched with egg yolk emulsion to determine phospholipase activity. These substrates contain phospholipids that are hydrolyzed by phospholipases.
   - **Control medium**: Add control with the substrate but without the fungal inoculum to exclude any background activity or contamination.
3. **Preparation of the inoculum**
   - Use mid-log fungal cultures (grown in rich media such as YEPD) for optimal enzyme secretion.

- Set the cell density to a standard level (*e.g.* 10^6–10^7 cells/mL) to ensure reproducibility.
- **Wash the cells:** Wash the fungal cells in sterile PBS to remove all residual nutrients before inoculation.

4. **Incubation conditions**
   - **Temperature:** Incubate the plates at a temperature between 30 °C and 37 °C, based on the optimal growth temperature of the fungal strain (30 °C for *Cryptococcus* sp. and 37 °C for *Candida* sp.)
   - **Incubation time:** Incubate the plates for 3–5 days (2–3 days for *Cryptococcus* sp. and 3–4 days for *Candida* sp.) to allow sufficient phospholipase activity.
   - **Aeration:** Provide adequate aeration or incubation in a humidified chamber to ensure optimal fungal growth.

17.4.2 Troubleshooting

1. **Problem**: **No precipitation zone**
   - **Possible causes**:
     - The fungal strain may not be producing phospholipase.
     - The egg yolk was not mixed properly into the SDA medium, preventing the reaction from taking place. The plates were not incubated at the correct temperature or under the correct conditions, inhibiting phospholipase production or activity.
   - **Solutions:**
     - **Confirm the fungal strain**: Ensure that the tested strain is known to produce phospholipase. Some fungi may not have phospholipase activity or require special environmental conditions to produce the enzyme.
     - **Ensure that the egg yolk is properly mixed**: When preparing egg yolk-enriched SDA, ensure the egg yolk is thoroughly mixed into the chilled SDA. High temperatures will cause the egg yolk to coagulate and impair its ability to interact with the phospholipase.
     - **Check the preparation of the medium**: Make sure the SDA has been freshly prepared and the egg yolk is not old or stored improperly, as this could interfere with the detection of phospholipase activity.
2. **Problem: Inconsistent precipitation zones**
   - **Possible causes:**
     - Inconsistentinoculation of fungal strains could lead to varying levels of enzyme production.
     - Inconsistent incubation temperatures can affect phospholipase expression, leading to irregular activity.
     - Improper mixing of egg yolk into the medium or insufficient cooling of the medium before the addition of egg yolk could lead to inconsistencies in the detection of phospholipase.

- **Solutions:**
  - **Standardize the inoculum size**: Ensure that the fungal inoculum is applied evenly and in a uniform volume on each plate. Using a uniform number of fungal cells or a sterile loop for each inoculation can help maintain consistency.
  - **Check the mixing of the egg yolk**: Make sure that the egg yolk is thoroughly mixed into the cooled SDA medium to avoid clumping or uneven distribution. Insufficient mixing can lead to inconsistent reactions and precipitation zones.
  - **Use an appropriate plating technique**: Ensure that the fungal colonies are appropriately spaced on the plate to avoid overlapping growth, which leads to inconsistent enzyme activity and precipitation zones.

3. **Problem: Poor incorporation of the egg yolk into the media**
   - **Possible causes:**
     - The egg yolk has coagulated due to the high temperatures, impairing its ability to interact with the phospholipase.
     - The egg yolk was added to a hot medium instead of a cooled medium, resulting in improper incorporation.
   - **Solutions:**
     - **Mix egg yolk into cooled medium**: After preparation, allow the SDA to cool to about 50–55 °C before adding the egg yolk. This prevents coagulation of the egg yolk, which can occur at higher temperatures and prevent effective phospholipase detection.
     - **Mix the egg yolk thoroughly**: Ensure that the egg yolk is evenly mixed into the cooled SDA. If necessary, use a sterile stirring rodor vortex to mix the egg yolk thoroughly into the medium.
4. **Problem: Cloudy agar plate**
   - **Possible causes:**
     - If the egg yolk is not handled properly, the lower, thicker layer of the egg yolk may be used, making the agar less transparent. This makes it difficult to see the precipitation zones around the fungal colonies.
   - **Solutions:**
     - After centrifugation, always use the upper, thinner layer of the egg yolk. The upper layer is more transparent and ensures that the plate remains clear to better see the precipitation zones around the fungal colonies. Avoid using the bottom layer as it is denser and can cloud the agar, obscuring the results.
5. **Problem: Variability of phospholipase activity**
   - **Possible causes:**
     - Some strains of fungi may express phospholipase under certain conditions, or enzyme activity may vary with fungal age or culture conditions.
   - **Solutions:**
     - **Confirm strain characteristics**: If using a mutant strain or different isolates, check if they express phospholipase activity under standard test conditions. Strains with mutations in phospholipase-related genes may show weak or no precipitation.

   - **Use age-matched cultures**: Ensure cultures are at the same growth stage. Older cultures may show increased phospholipase activity compared to younger cultures. Therefore, use freshly grown cultures of the same age to ensure consistency.
   - **Optimize the growth conditions**: Ensure that fungal strains are grown under optimal conditions to maximize phospholipase expression.
6. **Problem: Poor storage of the plates**
   - **Possible causes:**
     - Storage at excessively high temperatures or prolonged storage after incubation can lead to degradation of the phospholipase activity or the precipitation zone.
   - **Solutions:**
     - **Store the plates at 4 °C**: After the incubation period, keep the plates at 4 °C if necessary to preserve the precipitation zones and phospholipase activity. However, it is better to assess the precipitation zones immediately after incubation to obtain the clearest results.
     - **Limit the storage time of the plate**: Try to evaluate the results soon after incubation, as prolonged storage at low temperatures can sometimes obscure the precipitation zones.

Hydrophobicity Test for Virulence of Pathogenic Fungi

18

Abstract

Microbial adhesion to hydrocarbons (MATH) assay is widely used to evaluate the cell surface hydrophobicity (CSH) of fungi. CSH is an important feature that influences fungal adhesion to surfaces, biofilm formation and host tissue invasion, which are key factors in fungal pathogenicity (Danchik, Casadevall, Front Cell Infect Microbiol 10:594973, 2021). The microbial adhesion to hydrocarbons (MATH) assay evaluates the affinity of fungal cell surfaces for hydrophobic substances by measuring their ability to partition into a hydrocarbon layer, such as n-hexadecane, when mixed with an aqueous cell suspension. Hydrophobic cells preferentially migrate into the hydrocarbon phase, while hydrophilic cells remain in the aqueous phase. In this test, a fungal culture is mixed with n-hexadecane and then a phase separation is performed. The optical density (OD) of the aqueous phase (bottom layer) is measured after phase separation. A significant decrease in the OD of the aqueous phase indicates a high degree of hydrophobicity, as more cells adhere to the hydrocarbon layer, reducing the number of cells in the aqueous phase (de Miranda et al, J Appl Oral Sci 32:e20240154, 2024). Conversely, a minimal change in OD indicates low hydrophobicity. The degree of hydrophobicity reflects the chemical composition of the fungal cell surface, which is influenced by factors such as cell wall structure and the expression of surface proteins. The microbial adhesion to hydrocarbons (MATH) assay is crucial for studying fungal virulence, as hydrophobicity plays a critical role in the ability of fungi to adhere to host tissues and surfaces, which is a key step in the initiating infection (Kumari et al, Mycology 12:296, 2021). With this assay, we can compare the hydrophobicity of different fungal strains or investigate the effects of environmental conditions or genetic factors on the adhesion properties of fungi. Overall, the MATH assay is a valuable tool for improving our understanding of fungal pathogenicity and identifying potential therapeutic targets for

R. Pasrija, D. Kumari, *Protocols for Studying Pathogenic Fungi*,
https://doi.org/10.1007/978-981-96-7047-5_18

antifungal interventions. The protocol described describes the detailed steps of the MATH assay to measure CSH.

Keywords

MATH · Cell surface hydrophobicity · Hydrocarbon · Decrease · Aqueous layer

Introduction This chapter discusses detailed protocols for measuring fungal cell surface hydrophobicity (CSH) using the Microbial Adhesion to Hydrocarbons (MATH) assay. Topics covered include:

18.1. Materials Required
18.2. Preparations
18.3. Protocol
18.4. Tips and Troubleshooting

18.1 Materials Required

18.1.1 Fungal Strains and Growth Media

- **Fungal strains:** Pathogenic fungi such as *Candida* sp./*Cryptococcus* sp. or other relevant strains. Known hydrophobic and hydrophilic strains for control (optional).
- **Pre-culture media:** Yeast peptone dextrose broth (YEPD)

18.1.2 Reagents for the Test

- **Hydrocarbon layer:** Xylene, n-hexadecane, or toluene (used to test the hydrophobicity of fungal cells by assessing their adhesion to hydrophobic surfaces).
- **Phosphate-buffered saline (PBS):** For washing fungal cells and preparing suspensions.

18.1.3 Laboratory Consumables

- **Sterile tubes:** Glass or polypropylene tubes for mixing fungal suspension with hydrocarbon.
- **Pipettes and tips:** Sterile pipettes or micropipette tips for handling solutions.
- **Spectrophotometry cuvettes:** For optical density (OD) measurements, typically at 600 nm.

18.1.4 Equipment

- **Spectrophotometer:** For measuring the OD_{600} of the fungal suspension.
- **Vortex mixer:** For thoroughly mixing the fungal suspension and the hydrocarbon layer.
- **Incubator:** To grow fungal cultures at optimal temperatures (30 °C).

18.1.5 Controls

- **Positive control** (optional): A strain with high surface hydrophobicity.
- **Negative control** (optional): A strain with low or no hydrophobicity.

18.1.6 Environmental and Safety Materials

- **Personal Protective Equipment (PPE):** Lab coat, gloves, and goggles for safely handling hydrocarbons and pathogenic fungi.
- **Fume hood:** For working safely with volatile hydrocarbons such as xylene or toluene.
- **Biosafety cabinet:** For handling fungal cultures under aseptic conditions.

18.2 Preparations

18.2.1 Preparation of YEPD Media (See Chap. 2)

18.2.2 Preparation of PBS (See Chap. 3)

18.3 Protocol [1]

18.3.1 Preparation of Yeast Cells

- **Inoculate yeast cells**: Grow the yeast cells overnight in YEPD medium at 30 °C with constant shaking at 150 rpm.
- **Harvest cells**: Collect the yeast cells after 24 hours, by centrifuging the culture at 3000 × *g* for 5 min. Discard the supernatant.
- **Wash the cells**: Resuspend the cell pellet in PBS (pH 7.4) to remove residual growth medium. Centrifuge again and discard the supernatant. Repeat this washing step twice to ensure clean cells.
- **Adjust the cell concentration**: Resuspend the washed cells in PBS and adjust the OD_{600} to 0.2–0.4 using a spectrophotometer. This is the initial OD (A_0).

18.3.2 MATH Assay Procedure

- **Dispense cell suspension**: Transfer 3 mL of the cell suspension (OD_{600} 0.2–0.4) into a clean test tube or glass vial.
- **Add hydrocarbon**: Add 0.4 mL of n-hexadecane (or other hydrocarbons such as octane) to the tube containing the fungal suspension.
- **Vortex**: Vortex the mixture vigorously for 2 min to mix the aqueous and hydrocarbon phases thoroughly. This allows the fungal cells to interact with the hydrocarbon phase.
- **Phase separation**: After vortexing, allow the mixture to stand at room temperature or 30 °C for 15–30 min. During this time, the hydrocarbon and aqueous phases will separate, with the hydrocarbon forming the upper layer and the aqueous phase forming the lower layer. The more hydrophobic the fungal cells are, the more they will be distributed in the hydrocarbon phase. A high turbidity in the aqueous phase indicates a low hydrophobicity, as more cells remain in the aqueous phase.

18.3.3 Measuring the Optical Density

- **Collect aqueous phase**: After separating the phases, carefully remove 1 mL of the aqueous phase (bottom layer) from the test tube or vial to avoid contamination by the hydrocarbon layer.
- **Measure the OD_{600}**: Using a spectrophotometer, measure the OD_{600} of the aqueous phase and record this value as A_1.
- **Calculating the CSH**
  - To calculate the CSH percentage:

$$\left[1-\left(A_1 / A_0\right)\right] \times 100$$

- Where: A_0 is the initial OD600 of the cell suspension (before adding the hydrocarbon) and A_1 is the OD_{600} of the aqueous phase after phase separation.
- **Result interpretation:**
  - **High hydrophobicity**: If a significant decrease in OD_{600} is observed in the aqueous phase (*i.e.* A_1 is much lower than A_0), this indicates a high hydrophobicity of the fungal cells, which means that a greater proportion of the cells are adhered to the hydrocarbon layer (as shown in Fig. 18.1).
  - **Low hydrophobicity**: If the OD_{600} changes minimally between A_0 and A_1, the cells exhibit low hydrophobicity, meaning fewer cells have adhered to the hydrocarbon phase and most remain in the aqueous phase.
- **Notes**
  - Always use triplicate samples to ensure accuracy.
  - Ensure that all pipettes and vials are properly sterilized to avoid contamination.

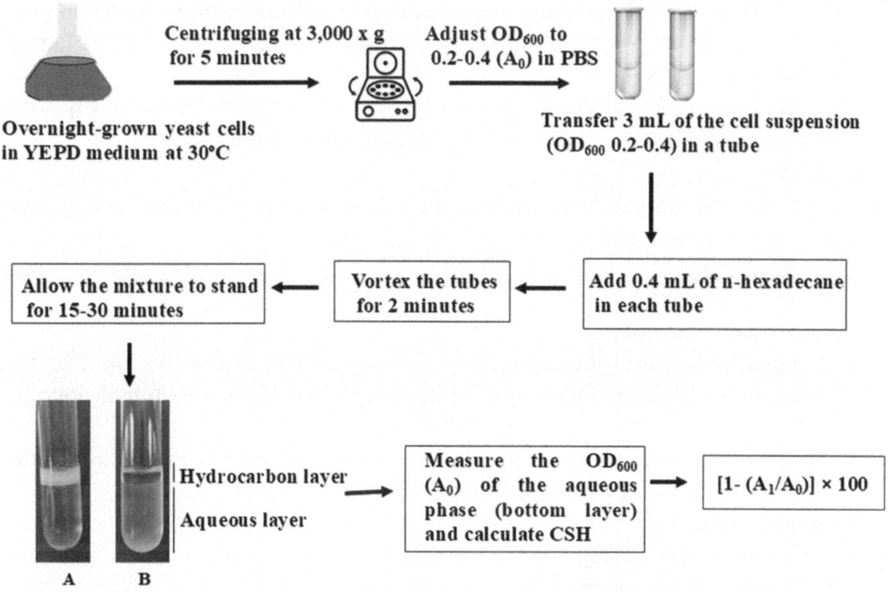

Fig. 18.1 Test of microbial adhesion to hydrocarbon of wild-type (WT) and mutant *Candida* cells in n-hexadecane: A indicates Wild-type *Candida* cells, while B indicates mutant *Candida* cells. Low turbidity in the aqueous phase with more cells in the hydrocarbon phase indicates higher hydrophobicity (WT), while high turbidity in the aqueous phase reflects lower hydrophobicity (mutant cells)

  - Mix samples consistently and uniformly using the same vortex time and intensity for each replicate.
- **Calculation example**
  - If the initial OD_{600} of the fungal suspension is 0.4 and the OD_{600} of the aqueous phase after mixing is 0.3, then the hydrophobicity of the cell surface is:

$$\text{Hydrophobicity}(\%) - (1 - 0.3/0.4) \times 100 = 25\%$$

18.4 Tips and Troubleshooting

18.4.1 Tips

1. **Selection of the fungal strain**
   - Use virulent fungal strains, such as *Candida* sp./*Cryptococcus* sp. which are known to have CSH.
2. **Preparation of the inoculum**
   - **Growth phase of the fungi**

  - Use fungal cells in the logarithmic phase to achieve optimal hydrophobicity, as they are more likely to express virulence factors, including changes in CSH.
  - **Incubation**: Grow the fungus in rich media (*e.g.* YEPD) for 18–24 h at the optimal temperature (30 °C for *Candida* and *Cryptococcus*).
- **Wash**
  - Wash the fungal cells with sterile PBS to remove residual media and metabolites that may interfere with the assay.
  - Resuspend the cells to the desired concentration, typically around OD_{600}– 0.2–0.4.

3. **Solvents and reagents**
   - **N-hexadecane or toluene** are common hydrophobic solvents used in the assay to evaluate the interaction of the fungal cell surface with hydrophobic compounds.
   - **Hydrophilic control:** Use a hydrophilic compound such as PBS to assess the interaction of the cells with a non-hydrophobic solvent.
   - **Assay methods**
   - **Microtiter plate assay:**
     - **Coating with solvent:** add n-hexadecane or toluene to sterile glass tubes.
     - **Inoculation:** Add a cell suspension to the glass tubes and allow the cells to interact with the solvent for a period of time.
     - **Measurement:** After incubation, measure the absorbance at 600 nm to determine the extent of cell adhesion to the hydrophobic solvent.

18.4.2 Troubleshooting

1. **Problem**: Poor or incomplete phase separation after vortexing may affect the results.
   - **Possible cause**:
     - Too much or too little vortexing, incorrect handling of the phases or insufficient time for phase separation can lead to incomplete separation of the aqueous and hydrocarbon phases.
   - **Solutions**:
     - **Allow the mixture to stand for a sufficient period of time**: After vortexing, allow the mixture to stand at room temperature/30 °C for 15–30 min to allow the phases to separate properly.
     - **Vortexing time**: Ensure to vortex the mixture for 1.5–2 min. Vortexing for too long can lead to emulsification, while vortexing for too short a time can result in the phases not mixing properly.
2. **Problem**: Fluctuating or inaccurate OD readings can lead to misleading results.
   - **Possible cause**:
     - Contaminated cuvettes, improper calibration, or incorrect measurement techniques can affect accurate OD readings.

- **Solutions:**
  - **Clean the cuvettes**: Always use clean cuvettes and wipe them with alcohol to avoid contamination by residues. Contamination in the cuvette can affect light absorption and lead to inconsistent readings.
  - **Calibrate the spectrophotometer**: Calibrate the spectrophotometer with PBS to ensure accurate readings. PBS should be used as a blank in the spectrophotometer to establish the baseline.
  - **Measure the OD_{600} in the appropriate phases**: Take the first measurements of the fungal cell suspension before adding the hydrocarbon phase and then take the measurements again after the phase separation. This will ensure that you capture any changes in cell concentration or hydrophobicity.
  - **Avoid contamination**: Ensure the sample is properly mixed and free of external contaminants, which could interfere with absorbance.

3. **Problem**: If the initial fungal cell suspension is too concentrated or dilute, it can affect the results of the MATH assay.
   - **Possible cause**:
     - Too high or too low an OD_{600} can lead to an inconsistent partitioning of fungal cells between the aqueous and hydrocarbon phases, skewing the results.
   - **Solutions:**
     - **Keep the OD_{600} between 0.2 and 0.4**: Ensure that the OD_{600} of the cell suspension is between 0.2 and 0.4 before adding it to the test. This is the optimal concentration for evaluating the hydrophobicity of fungi.
     - **Wash the cells thoroughly**: Wash the fungal cells thoroughly to remove any residual medium or nutrients that may interfere with the test. Residual medium can affect the partitioning of the cells and lead to inaccurate results.
     - **Check for clumping**: Ensure that the fungal cells are well dispersed and not clumped together, as clumping can affect the separation process and lead to inaccurate results.
4. **Problem**: Degradation of the hydrocarbon (*e.g.* n-hexadecane) can affect the partitioning of the fungal cells in the hydrocarbon phase.
   - **Possible cause**:
     - The use of old or improperly stored hydrocarbons can lead to decomposition or evaporation and affect the test result.
   - **Solution**
     - **Use fresh n-hexadecane**: Always use fresh n-hexadecane (or another suitable hydrocarbon) for each test. Store the hydrocarbon properly in a tightly sealed container to avoid evaporation and degradation.
     - **Proper storage**: Store the hydrocarbon in a cool, dry place, away from direct sunlight, to maintain its quality. If using the hydrocarbon over multiple experiments, ensure it is stored under appropriate conditions (*e.g.* tightly sealed). If you use the hydrocarbon over several experiments, ensure it is stored under suitable conditions (*e.g.* tightly sealed).

    - **Avoid contamination**: Ensure the hydrocarbon is not contaminated with water or other solvents that will interfere with phase separation.

5. **Problem**: If too few cells are recovered in the hydrocarbon phase, this could indicate a problem with cell binding or phase separation.
    - **Possible cause**:
        - Improper vortexing or phase separation or low hydrophobicity of the fungal strain tested, may result in fewer cells being partitioned into the hydrocarbon phase.
    - **Solutions**:
        - **Increase the vortex intensity**: If you do not see enough cells in the hydrocarbon phase, try to increase the vortex intensity, but be careful not to vortex too much.
        - **Check fungal strain**: Ensure the tested strain can disperse in the hydrocarbon phase. Some strains may be less hydrophobic and, therefore, have a lower recovery in the hydrocarbon phase. If necessary, test several strains to ensure consistency.

References

1. Vij R, Danchik C, Crawford C et al (2020) Variation in cell surface hydrophobicity among *Cryptococcus neoformans* strains influences interactions with amoebas. mSphere 5. https://doi.org/10.1128/msphere.00310-20

Adhesion of Fungal Cells to Buccal Epithelial Cells

19

Abstract

The ability of fungi to adhere to host epithelial cells is a crucial step in the initiation of infection and pathogenesis of various fungal diseases (Martin et al, Future Med Chem 13:313–334, 2021). To study this adhesion process, a detailed model of fungal epithelial cell adhesion mimics the interaction between fungal cells and host tissue. The adhesion of pathogens is usually facilitated by microbial proteins, so-called adhesins, which bind to cell surface biomolecules such as protein receptors and glycans. In particular, the oral cavity provides surfaces to which fungal cells can adhere, such as buccal epithelial cells (BECs) (Martin et al, Eur J Med Chem 160:82–93, 2018). This model is based on the principle that fungal cell surface components, such as glycoproteins and adhesins, interact with specific receptors on the epithelial cell surface to enable adhesion. The extent of adhesion provides valuable information about the virulence of the fungal strain, the ability of the fungus to colonize host tissue and its potential to cause infection (Yang et al, Microbiol Res 169(11):803–810, 2014). This assay provides a simple but effective method to study host-pathogen interactions and can be used to test the efficacy of potential antifungal agents that interfere with fungal adhesion (Moyes et al, Virulence 6(4):338–346. 2015). Understanding these adhesion mechanisms is crucial for the development of strategies for the prevention and treatment of fungal infections. In this chapter, we present a detailed protocol for assessing fungal adhesion in which fungal cells are co-incubated with epithelial cells to allow cell-surface interactions. After a specific incubation period, the non-adherent fungal cells are washed away, and the adherent cells are visualized and quantified using staining techniques, such as crystal violet, which provide clear and reproducible results. These visual assays provide a straightforward method for assessing the extent of fungal adhesion to host cells *in vitro*.

R. Pasrija, D. Kumari, *Protocols for Studying Pathogenic Fungi*,
https://doi.org/10.1007/978-981-96-7047-5_19

Keywords

Infection · Fungal cells · Adhesin · Receptors · Oral cavity · Epithelial cells

Introduction This chapter deals with detailed protocols for the adhesion of fungal cells to epithelial cells of the cheek. Topics covered include:

19.1. Materials Required
19.2. Preparation of Media and Buffers
19.3. Protocol for Adhesion of Fungal Cells to Buccal Epithelial Cells
19.4. Tips and Troubleshooting

19.1 Materials Required

19.1.1 Biological Materials

- **Fungal strains:** Pathogenic fungi (*e.g. Candida/Cryptococcus* sp.). Use wild-type and mutant strains when studying differences in adhesion.
- **Epithelial cells:** Freshly harvested human buccal epithelial cells from consenting healthy donors. Alternatively, immortalized epithelial cell lines can be used for *in vitro* models.
- **Growth media for fungi:** Yeast peptone dextrose broth (YEPD), Roswell Park Memorial Institute (RPMI-1640) or Sabouraud dextrose broth (SDB):
- **Buffered solution for cell harvesting:** phosphate-buffered saline (PBS) (For washing buccal and fungal cells).

19.1.2 Staining Agents and Reagents

- **Stains for visualization**
  - **Trypan blue/crystal violet:** Staining of adherent fungi and epithelial cells.
  - **Calcofluor white (CFW) (optional):** For fluorescent staining of fungal cells.
- **Fixative**
  - **Methanol, nail paint, or formaldehyde:** For fixing epithelial cells on slides.
  - **Embedding medium (optional):** For microscopic analysis (*e.g.* Fluoromount-G for fluorescence microscopy).
- **Solution of an antifungal agent (optional):** When testing the effect of treatments on fungal adherence.

19.1.3 Laboratory Consumables

- **Sterile tubes:** For fungal cell suspensions and epithelial cell collection.
- **Glass slides or microtiter plates:** For setting up the adhesion test.
- **Coverslips:** For overlaying stained samples for microscopic analysis.
- **Sterile swabs or scrapers:** For collecting buccal epithelial cells.
- **Micropipettes and sterile tips:** For accurate handling of liquids.

19.1.4 Instruments

- **Microscope:** Light microscope or fluorescence microscope, depending on the staining used.
- **Centrifuge:** For pelleting fungal cells and epithelial cells.
- **Incubator:** For holding the fungal cultures at 30–37 °C.
- **Hemocytometer or spectrophotometer:** For standardizing the concentrations of fungal and epithelial cells.

19.2 Preparation of Media and Buffers

19.2.1 Preparation of YEPD Media (See Chap. 2)

19.2.2 Preparation of RPMI-1640 (See Chap. 2)

19.2.3 Preparation of PBS (See Chap. 3)

19.2.4 Preparation of Spider Media (See Chap. 8)

19.3 Protocol [1]

- **Preparation of the fungal cells**
  - **Grow the fungal cells:** Grow the yeast cells overnight in YEPD medium (or other suitable medium) for 24 h at 30 °C with shaking at 150 rpm to promote growth.
  - **Harvest fungal cells:** Centrifuge the culture after incubation at 3000×*g* for 5 min to pellet the cells. Wash the fungal pellet twice with PBS to remove the residual medium. Resuspend the fungal cells in Spider medium or RPMI-1640 medium at a concentration of 10^7 cells/mL.
- **Collection of buccal epithelial cells from the subject**

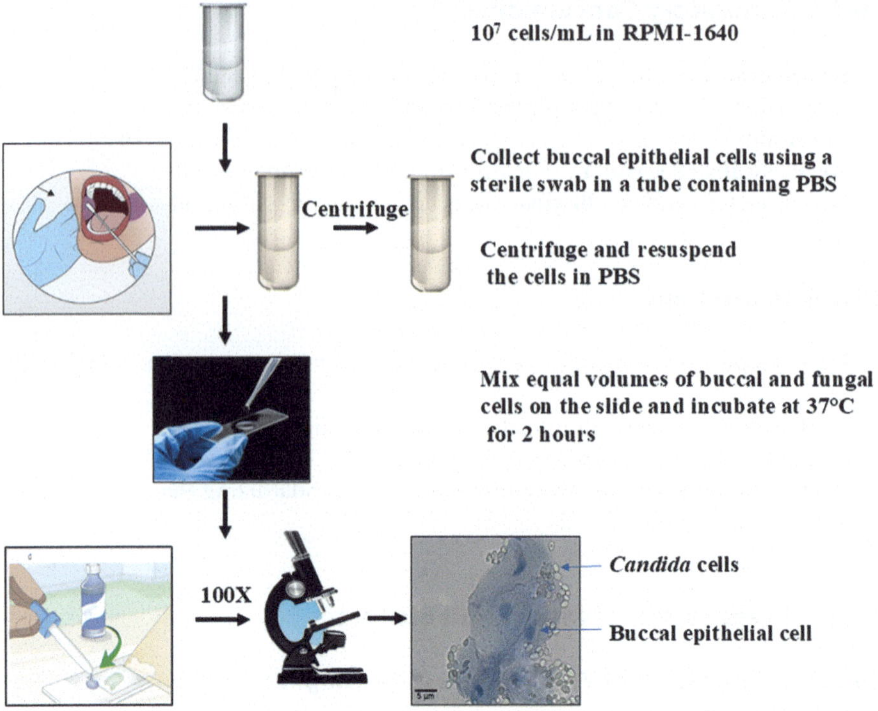

Fig. 19.1 Flow chart illustrating the adhesion of *Candida* cells to epithelial cells of the cheek, observed under the microscope at 100× magnification

  - **Collect cells from volunteer:** Using a sterile cotton swab, collect buccal epithelial cells from the inside of a subject's cheek (as shown in Fig. 19.1). Alternatively, the subject can rinse their mouth with sterile saline or PBS to collect cells.
  - **Centrifuge the epithelial cells:** Centrifuge the collected cell suspension at 300–500×*g* for 5 min to pellet the epithelial cells.
- **Resuspend epithelial cells**
  - After centrifugation, resuspend the epithelial cell pellet in PBS to achieve the required concentration for co-incubation with fungal cells.
- **Adhesion Assay**
  - **Prepare the co-incubation:** Mix equal volumes of the fungal and buccal epithelial cell suspension on a sterile glass slide/microtiter plate. This ensures that the fungal and epithelial cells are well mixed for the co-incubation.
  - **Incubate the slide:** Place the slide/plate in a 37 °C incubator for 2 h to allow the fungal cells to adhere to the epithelial cells. Gently shake the slide occasionally during incubation to ensure even distribution of the cells. This step allows time for adhesion between the fungal cells and the epithelial cells.
- **Washing**

  - After incubation, gently wash the slide with PBS to remove non-adherent fungal cells. Ensure the washing process is carried out carefully so that the epithelial cells do not shift.
- **Staining and microscopy**
  - **Stain the cells**: Stain the slide with a suitable staining solution (*e.g.* 0.4% trypan blue or crystal violet) to visualize fungal and epithelial cells under the microscope.
  - **Trypan blue:** Add 2–3 drops of 0.4% trypan blue/crystal violet to the slide, mix gently, and incubate for 5 min.
  - **Fix the slide**
    - Prepare 4% paraformaldehyde in PBS (a commonly used fixative).
    - Immerse the slide in the 4% formaldehyde solution for 10–15 min at room temperature. This preserves the cells by stabilizing their structure and preventing their degradation.
  - **Washing after fixation**
    - After the formaldehyde fixation, rinse the slide thoroughly with PBS (2–3 washes of 5 min each) to remove all traces of fixative.
  - **Sealing with nail varnish** (optional)
    - After staining with trypan blue and washing with PBS, dry the slide by gently tapping it on a paper towel to remove any residual PBS or moisture.
    - Take nail polish and apply a thin layer around the edges of the coverslip. This will act as a sealant to keep the cover glass in place. Do not apply the nail polish directly to the cells to not impair the view.
    - Allow the nail polish to dry for about 5–7 min to fix the coverslip in place.
  - **Microscopic visualization**: Examine the slide under a microscope at 100× magnification to observe the adhesion of the fungal cells to the epithelial cells (see Fig. 19.1). If quantification is required, count the number of fungal cells adhering to each epithelial cell.

19.4 Tips and Troubleshooting

19.4.1 Tips

1. **Selection of epithelial cells**
   - Ensure that the epithelial cells for the test are healthy and actively growing.
2. **Fungal strain and growth conditions**
   - **Virulent fungal strains** (*e.g.*, *Candida* sp., *Cryptococcus* sp.) should be used for the adhesion test.
   - Use overnight-grown fungal cultures, which generally have the strongest adhesion due to their cell surface properties.

- **Preparation of the fungal inoculum**: Ensure that the fungal cells are properly washed in PBS to remove any residual growth medium that may interfere with adhesion. Resuspend the fungal cells to a concentration (*e.g.* 10^6–10^7 cells/mL) to ensure reproducibility.

3. **Co-culture conditions**
   - **Confluence of the epithelial cells**: Ensure that epithelial cells are 80-100% confluent before fungal inoculation. Confluence ensures a monolayer that provides an optimal substrate for adhesion.
   - **Duration of adhesion**: Allow an adequate incubation time (usually 2–3 h) for the fungal cells to adhere to the epithelial cell monolayer. A longer incubation time may be required for certain types of fungi. **Incubation conditions**: Maintain the co-culture at 37 °C to simulate physiological conditions.
4. **Protocol for the adhesion test**
   - **Inoculation:** Add the fungal suspension (at a suitable density) to the monolayer of epithelial cells. Gently swirl or incubate the plates to ensure even distribution of fungal cells over the epithelial surface.
   - **Washing after incubation:** Wash the co-culture several times with sterile PBS to remove non-adherent fungal cells. Be careful when washing to avoid displacing adherent fungal cells.
   - **Fungal detection:** Stain the fungal cells with trypan blue/crystal violet to make them visible under the light microscope. However, Calcofluor White (CFW) can be used for fluorescent microscopes. After removing the non-adherent cells, quantify the adherent fungal cells either by microscopic counting or by CFU assay (optional).

19.4.2 Troubleshooting

1. **Problem:** Low or inconsistent fungal attachment to epithelial cells.
   - **Possible cause**: If the fungal cells are old, poorly cultured, or dead, their adhesion properties may be significantly compromised.
   - **Solutions**:
     - **Use fresh fungal cultures**: Always start the test with fresh, actively growing ones. Ensure that the fungal cells are in the correct growth phase (usually the exponential phase) to achieve optimal adhesion properties.
     - **Confirm fungal viability**: Before performing the adhesion test, check the viability of the fungal culture with a viability test (*e.g.* with a dye such as trypan blue) to confirm that the cells are alive.
2. **Problem: Weak or no fungal adhesion to epithelial cells [2]**
   - **Possible cause**: Insufficient incubation time for the fungal cells to adhere to the epithelial cells or insufficient concentration of fungal cells.
   - **Solutions**:
     - **Adjust the incubation time**: The optimal incubation time for fungal adhesion is 2 h. If adhesion weakens, increase the incubation time by 30–60 min.

  - **Increase the fungal concentration**: Try to increase the concentration of fungal cells in the assay. Use a higher fungal cell density to avoid no/limited fungal cell visualization or if adhesion appears weak.

3. **Problem: Loss of epithelial cells during washing or inaccurate fungal adhesion results**
   - **Possible cause**: Vigorous or overzealous washing can cause the epithelial cells to detach from the slide, resulting in inaccurate or unreliable adhesion measurements.
   - **Solutions**:
     - **Wash gently**: When washing the slides to remove unattached fungal cells, do so gently with a low-pressure stream of PBS. Ensure the epithelial cells are not disturbed or detached by the washing process.
     - **Use a minimal volume of PBS**: Use smaller, controlled amounts of PBS to wash the slide rather than using large volumes of PBS. This reduces the risk of detachment of epithelial cells.
     - **Optimize the washing steps**: Wash the slides 2–3 times in PBS to ensure proper removal of non-adherent fungal cells. However, avoid excessive washing, which could lead to the loss of epithelial cells.
4. **Problem: Difficulty visualizing fungal cells or poor contrast between fungal cells and epithelial cells**
   - **Possible cause**: Staining problems, such as insufficient dye uptake, short staining times, or ineffective staining agents, can hinder the visualization of fungal cells on epithelial cell surfaces.
   - **Solutions**:
     - **Extend the staining time**: If the fungal cells are difficult to detect, extend the staining time. Allow the staining agents to interact with the fungal cells for 5–15 min.
     - **Ensure uniform staining**: Carefully wash the slides after staining to remove excess dye and improve contrast. Check under the microscope that the fungal cells are visible.
5. **Problem**: Difficulty in accurately quantifying the number of fungal cells adhering to the epithelial cells.
   - **Possible cause**: Inconsistent counting methods, inadequate controls, or poor visualization of the fungal cells can lead to inaccurate quantification of adhesion.
   - **Solutions**:
     - **Perform repeat tests**: Always perform the adhesion test in replicates (at least three biological replicates) to account for variability and ensure reliable results.
     - **Manual counting or image analysis**: For quantification, manually count the number of fungal cells adhering to each epithelial cell under the microscope or use image analysis software to count the fungal cells more accurately.

    - **Standardize the counting criteria**: Establish a standardized criterion for "attachment" (*e.g.* fungal cells directly touching the epithelial cells) and ensure that all replicates follow the same counting protocol.
    - **Use appropriate magnification**: Use magnification of at least 100× to visualize adhesion at the single-cell level and make sure that both the fungal cells and the epithelial cells are distinguishable.

6. **Problem**: Inconsistent or poor quality of epithelial cell adhesion or overgrowth.
   - **Possible cause**: If the epithelial cells are not healthy, they may not adhere properly to the slide, or if the cell density is too low, fungal adhesion may be difficult to assess.
   - **Solutions**:
     - **Maintain healthy epithelial cells**: Ensure that the epithelial cells are healthy. The cells should be freshly isolated or cultured under optimal conditions.
     - **Optimize the epithelial cell density**: Ensure that the density of epithelial cells on the slide is appropriate. If the cells are too sparse, you may not have enough for accurate adhesion analysis. If they are too dense, the cells may overlap and make it difficult to assess individual adhesion events.

References

1. Pasrija R, Kumari D, Wadhwan S, Chhillar A, Mittal V (2023) The *Zanthoxylum armatum* fruit's oil exterminates *Candida* cells by inhibiting ergosterol biosynthesis without generating reactive oxygen species. Int Microbiol 27:1–12. https://doi.org/10.1007/s10123-023-00401
2. Segal E, Gottlieb S, Altboum Z, Gov Y, Berdicevsky I (1997) Adhesion of *Candida albicans* to epithelial cells—effect of nikkomycin. Mycoses 40(1–2):33–39. https://doi.org/10.1111/j.1439-0507.1997.tb00168.x

Sterols Extraction, Analysis and Quantification in Pathogenic Fungi

20

Abstract

Sterol extraction and analysis are essential techniques for studying pathogenic fungi, as sterols, particularly ergosterol, play a pivotal role in fungal pathogenesis. Ergosterol is a key component of the fungal cell membrane, crucial for maintaining its integrity and fluidity, which are essential for cell growth and survival (Choy et al, mBio 14:e01353, 2023). This sterol enables important functions such as nutrient uptake, signal transduction, and protection against environmental stressors. Disruption of ergosterol biosynthesis has become a primary target for antifungal therapies, as it compromises membrane function, ultimately impairing fungal virulence and reducing the ability of the pathogen to infect its host. The method described in this chapter focuses on extracting sterols from fungal cells and quantifying their content to assess fungal membrane composition. In this process, fungal cells are first cultured overnight and subjected to alkaline hydrolysis using alcoholic potassium hydroxide (KOH), which breaks down cellular structures and releases the sterols from the cell membranes (Arthington-Skaggs et al, J Clin Microbiol 37(10):3332–3337, 1999). The sterols are subsequently extracted using petroleum ether, a non-polar solvent that effectively dissolves lipophilic compounds, including sterols. The extraction process ensures the separation of sterols from other cell debris and impurities. Once extracted, the sterol content is quantified using UV-VIS spectrophotometry (Pasrija et al, J Antimicrob Chemother 55:905–913, 2005). The absorbance of the sample is measured at specific wavelengths that correspond to the characteristic absorbance peaks of these sterols. This technique allows for the quantification of ergosterol and related sterols, which are integral components of fungal cell membranes and play a critical role in membrane fluidity and function (Pasrija et al, Int Microbiol 27:1–12, 2023). The method provides a reliable, reproducible approach for quantifying ergosterol, offering valuable insights into fungal membrane composition and the effects of antifungal treatments. By assessing the ste-

R. Pasrija, D. Kumari, *Protocols for Studying Pathogenic Fungi*,
https://doi.org/10.1007/978-981-96-7047-5_20

rol profile, researchers can gain a better understanding of how antifungal drugs interact with the fungal membrane and potentially lead to resistance. The chapter outlines a straightforward yet comprehensive protocol that can be used to study ergosterol biosynthesis in pathogenic fungi, providing a basis for the development of new antifungal strategies.

Keywords

Sterol · Ergosterol · YEPD · Hydrolysis · Petroleum ether · Fungal membrane

Introduction This chapter will cover detailed protocols for sterols extraction, analysis and quantification in pathogenic fungi. The topics addressed include:

20.1. Materials Required
20.2. Preparation of Media and Buffers
20.3. Protocol
20.4. Tips and Troubleshooting

20.1 Material Required

20.1.1 Biological Materials

- **Fungal strains:** Pathogenic fungi of interest. Use wild-type and mutant strains if assessing differences in sterol content.
- **Growth media:** Yeast Peptone Dextrose (YEPD) broth or Sabouraud Dextrose Broth (SDB) (For culturing fungi to sufficient biomass) or minimal media (optional): (For nutrient-limited conditions if sterol biosynthesis is being studied).

20.1.2 Chemicals and Reagents

- **Organic solvents:** Methanol/petroleum ether or chloroform. All solvents should be of HPLC grade.
- **Saponification reagents:** Potassium hydroxide (KOH) (For preparing the saponification solution in methanol)
- **Acid for neutralization:** Hydrochloric acid (HCl) (For neutralizing the alkaline solution after saponification).
- **Sterol standards:** Ergosterol, cholesterol (For creating standard curves during quantification).
- **Water:** Deionized or ultrapure water for reagent preparation.

20.1.3 Equipment

- **Centrifuge:** For pelleting fungal biomass.
- **Lyophilizer or vacuum concentrator**: To dry fungal cells before extraction.
- **Homogenizer or bead mill:** For disrupting fungal cells and ensuring efficient sterol extraction.
- **Water bath or heating block:** To maintain the required temperature during saponification.
- **Spectrophotometer:** For preliminary sterol detection using UV absorbance (*e.g.* at 280 nm for ergosterol).

20.1.4 Safety Materials

- **Personal protective equipment:** Lab coat, gloves, goggles, and face mask for handling organic solvents and caustic reagents.
- **Fume hood:** For safe handling of chloroform, methanol, and other volatile solvents.

20.2 Preparation of Media and Buffers

20.2.1 Preparation of YEPD Media (See Chap. 2)

20.2.2 Preparation of 25% Alcoholic KOH Solution

- **Materials required**
- KOH pellets, ethanol (absolute or 95% ethanol), beaker or glass container, stirring rod or magnetic stirrer, balance (for measuring KOH), pH meter or pH indicator strips (optional, for checking pH), distilled water (for rinsing), *etc.*
- **Procedure**
  - **Calculate the amount of KOH:** To prepare 100 mL of 25% alcoholic KOH solution, you will need 25 g of KOH.
  - **Weigh the KOH:** Use a balance to weigh the required amount of KOH pellets (*e.g.* 25 g for 100 mL of solution). Transfer the KOH to a clean, dry beaker or glass container.
  - **Add ethanol:** Pour ethanol (absolute or 95% ethanol) into the beaker with the KOH. The total volume of ethanol should be adjusted to ensure that the final solution is 25% KOH by weight. For 100 mL of the final solution, add 75 mL of ethanol.
  - **Dissolve the KOH:** Stir the mixture using a stirring rod or place the beaker on a magnetic stirrer. Ensure that all the KOH pellets dissolve completely in the ethanol. This may take several minutes. If any KOH is undissolved, continue stirring or gently heating the mixture (under controlled conditions) to facilitate dissolution.

- **Check and Adjust pH (Optional):** Although the solution is likely to be highly alkaline, you can check the pH using a pH meter or pH indicator strips. It should fall between 12 and 14. If necessary, add small amounts of ethanol to adjust the volume or concentration.
- **Storage:** Transfer the prepared 25% alcoholic KOH solution to a sealed, labelled container. Store the solution in a cool, dry place away from direct sunlight.

20.3 Protocol

- **Inoculation and growth:** Begin by inoculating 50 mL of sterile YEPD medium with an overnight culture of *C. albicans*. Incubate the inoculated culture at 30 °C with shaking at 150 rpm for 16 h to achieve robust fungal growth.
- **Harvesting cells:** After 16 h of incubation, centrifuge the culture at 4000×*g* for 5 min to collect the fungal cells. Discard the supernatant and wash the cell pellet twice with distilled water to remove any residual medium. After washing, record the wet weight of the cell pellet, as this will be used for further calculations.
- **Sterol extraction:** Add 3 mL of 25% alcoholic KOH to the cell pellet in each vial. Vigorously shake the mixture for 1 min to ensure proper mixing and cell lysis. Transfer the vials to a water bath at 85 °C for 1 h to allow the KOH to hydrolyze cellular lipids and release the sterols. After the incubation period, remove the vials and let them cool to room temperature before proceeding with the extraction.
- **Extraction with petroleum ether:** Add 4 mL of petroleum ether to each vial containing the KOH-treated cell suspension. Shake the vials vigorously for 2–3 min to extract the sterols into the petroleum ether phase. Carefully transfer the upper petroleum ether layer (containing the sterols) into clean glass vials, ensuring no contamination from the aqueous phase. Store the petroleum ether extracts at −20 °C for up to 24 h before proceeding with further analysis
- **Sterol analysis:** For UV–VIS spectrophotometric analysis, dilute the petroleum ether samples fivefold using fresh petroleum ether. Measure the absorbance spectra of the samples between 200 nm and 320 nm using a UV–VIS double-beam spectrophotometer. Record the absorbance at the relevant wavelengths, specifically at 281.5 nm (for ergosterol) and 230 nm (for 24(28)-dehydroergosterol) (as shown in Fig. 20.1).
- **Calculation of ergosterol and dehydroergosterol:** To calculate the percentage content of ergosterol and 24(28)-dehydroergosterol, use the following formulas:

$$\%\text{Ergosterol} + \%24(28) - \text{DHE} = \left[\left(A_{281.5} / 290\right) * F\right] / \text{pellet weight}$$

$$\%24(28) - \text{DHE} = \left[\left(A_{230} / 518\right) * F\right] / \text{pellet weight}$$

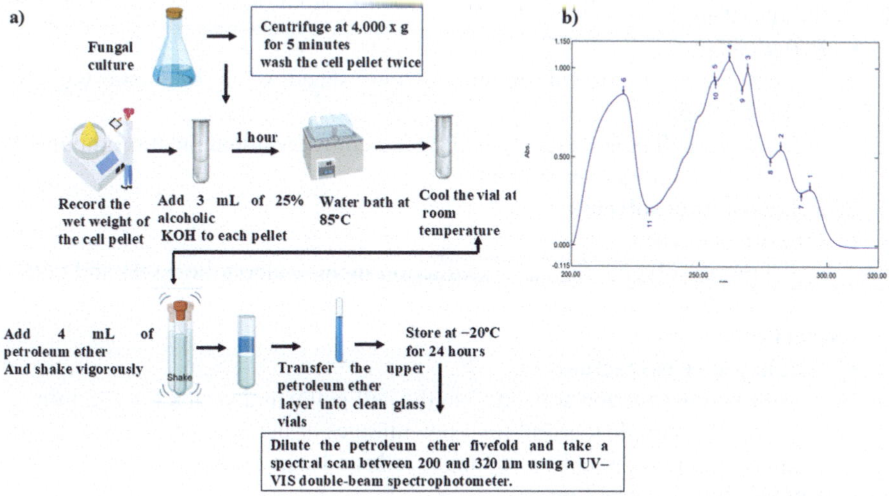

Fig. 20.1 (**a**) Protocol showing sterol extraction in fungi and (**b**) Graph showing sterol analysis

$$\%\text{Ergosterol} = \left[\%\text{Ergosterol} + \%24(28) - \text{DHE}\right] - \%24(28) - \text{DHE}$$

- where, $A_{281.5}$ and A_{230} are the absorbances at 281.5 nm and 230 nm, respectively. F is the dilution factor for the petroleum ether. 290 and 518 are the specific E values for crystalline ergosterol and 24(28)-DHE, respectively. Pellet weight is the weight of the fungal pellet used in the extraction [1].
- **Safety notes**
  - Petroleum ether is highly flammable and should be handled in a well-ventilated area, away from open flames or heat sources.
  - KOH is caustic, so always wear appropriate personal protective equipment, including gloves, lab coat, and safety goggles.
  - Sterile techniques should be followed to avoid contamination of fungal cultures and extracts.

20.4 Tips and Troubleshooting

20.4.1 Tips

1. **Sample preparation**
   - **Fungal growth**
     - Grow the fungal cultures overnight in appropriate growth media (*e.g.* YEPD, SDB).
     - Ensure cultures are in the log phase for a consistent sterol content.
     - Harvest approximately 10^7–10^8 cells per sample for extraction to ensure sufficient sterol yield.

2. **Cell harvesting**
   - **Pellet cells:**
     - Centrifuge the fungal culture at low speed (*e.g.* 3000–5000×g) for 10–15 min.
     - Wash the cell pellet with sterile PBS to remove residual media components and any external contaminants.
3. **Sterol extraction solvents**
   - **Organic solvents**
     - Common solvents for sterol extraction include alcoholic KOH and petroleum ether.
4. **Extraction process**
   - **Single-phase extraction**
     - Add your extraction solvent(s) to the cell pellet and vortex for 1–2 min.
     - Allow the sample to stand for a few minutes or use sonication (at 4 °C) for more efficient disruption of fungal cells and sterol release.
   - **Liquid-liquid extraction**
     - After mixing the cells with the solvent, add water and centrifuge to separate the organic phase containing sterols.
     - Carefully collect the organic layer and repeat the extraction if needed for complete recovery of sterols.
   - **Drying and resuspension**
     - Evaporate the solvent under nitrogen or using a rotary evaporator.
     - Resuspend the dried extract in a small volume of methanol/petroleum ether or another solvent suitable for the analysis method (*e.g.* chloroform for GC analysis).
5. **Storage of extracts**
   - Store at −20 °C for long-term storage to prevent degradation of sterols. Avoid repeated freeze-thaw cycles.
6. **Sterol standards**
   - Use pure sterol standards (*e.g.* ergosterol, cholesterol, lanosterol) for calibration and identification.
   - Ensure proper standard preparation and storage to avoid contamination or degradation.
7. **Quantification**
   - **External standard curve:** Generate a calibration curve using known concentrations of sterol standards, typically ranging from 0.1 to 10 μg/mL, depending on expected sterol concentration.
   - **Internal standard method:** Quantify sterols relative to an internal standard (such as cholesterol) to account for losses during extraction.

20.4.2 Troubleshooting

1. **Problem**: **Poor sterol extraction or low yields of sterols.**
   - **Possible cause**:

    - Inadequate mixing during the addition of KOH or petroleum ether may result in incomplete cell lysis or inefficient extraction of sterols.
- **Solution**:
    - **Ensure thorough mixing**: When adding KOH to lyse fungal cells, and petroleum ether for extraction, mix the solution thoroughly using vortexing or gentle shaking to ensure even distribution of solvent and lysis reagents. If the mixture is not homogeneous, the sterols will not be efficiently extracted.
    - **Check the cell pellet**: After lysis, ensure that the fungal cells have been fully disrupted. Any remaining clumps of cells can reduce extraction efficiency. Consider using a sonicator or mechanical disruption if necessary to fully lyse the cells.
    - **Longer incubation**: If cell lysis is incomplete, incubate for an additional 10–15 min with gentle agitation to promote better extraction.

2. **Problem**: **Contamination of the sterol extract with the aqueous phase, leading to impure samples and inaccurate readings**
    - **Possible cause**:
        - Improper separation of the petroleum ether layer from the aqueous phase can lead to cross-contamination of the two layers.
    - **Solution**:
        - **Use a separating funnel or pipette carefully**: When separating the petroleum ether layer, ensure that you remove it gently without disturbing the aqueous phase. Tilt the extraction container to encourage clear separation.
        - **Avoid disturbing the aqueous phase**: If using a separating funnel, carefully drain the aqueous phase from the bottom and collect only the petroleum ether layer at the top. If using a pipette, slowly aspirate the upper phase without touching the interface.
        - **Repeat extraction if necessary**: If you suspect contamination, perform a second extraction with fresh petroleum ether to improve purity.
3. **Problem: Reduced extraction efficiency and lower yields of sterols**
    - **Possible cause**:
        - Petroleum ether can evaporate quickly or degrade over time, especially if stored improperly. Stale or evaporated solvent reduces extraction efficiency.
    - **Solution**:
        - **Always use fresh petroleum ether**: For best results, ensure that the petroleum ether is freshly prepared for each extraction. Store the petroleum ether in tightly sealed containers and avoid exposing it to air for long periods.
        - **Check solvent condition**: If the solvent appears to have evaporated or smells off, discard it and use a fresh batch.
        - **Keep solvent tightly sealed**: When not in use, store petroleum ether in a well-sealed container to minimize evaporation.

4. **Problem**: **Inaccurate absorbance readings lead to incorrect sterol quantification**
   - **Possible cause**:
     - An improperly calibrated spectrophotometer can lead to incorrect readings, resulting in erroneous sterol concentration measurements.
   - **Solution**:
     - **Calibrate the spectrophotometer**: Before taking any measurements, ensure that the spectrophotometer is calibrated with an appropriate blank solution, typically the solvent or buffer used in the extraction (*e.g.*, petroleum ether or ethanol).
     - **Use appropriate wavelength**: Check that you are measuring absorbance at the correct wavelength for sterol quantification, usually around 282 nm for sterol detection in spectrophotometric assays.
     - **Recalibrate if necessary**: If you notice inconsistent readings or suspect calibration issues, recalibrate the spectrophotometer using fresh blank and standard solutions.
5. **Problem: Degradation of sterols due to temperature sensitivity**
   - **Possible cause**:
     - Sterols are sensitive to heat and can degrade or isomerize if left at room temperature for prolonged periods.
   - **Solution**:
     - **Keep samples on ice**: During all steps of handling and storing the sterol extracts, keep the samples on ice to minimize any degradation of sterols. This includes extraction, phase separation, and before quantification.
     - **Store extracts at − 20 °C or lower**: After extraction, store the sterol extracts at −20 °C or − 80 °C if they are not to be used immediately. This will prevent the breakdown of sterols.
6. **Problem: Emulsions can form during extraction, making it difficult to separate the two phases**
   - **Possible cause**:
     - Improper solvent mixing or inadequate settling time can cause emulsions, which hinder phase separation.
   - **Solution**:
   - **Let phases settle completely**: After mixing and allowing the phases to separate, let the mixture settle undisturbed for 10–20 min. If emulsions persist, consider adding a small amount of salt solution (e.g., NaCl) to facilitate phase separation.
   - **Use a mild vortex or shake**: Avoid over-agitating during mixing, as this can create emulsions. Gentle mixing should suffice to achieve proper extraction.

References

1. Singh A, MacKenzie A, Girnun G, Del Poeta M (2017) Analysis of sphingolipids, sterols, and phospholipids in human pathogenic *Cryptococcus* strains. J Lipid Res 58(10):2017–2036. https://doi.org/10.1194/jlr.M078600

Assessing Mitochondrial Fitness by Assessing Microbial Growth on Non-fermentable Carbon Sources

21

Abstract

Mitochondria are critical organelles for energy production in eukaryotic cells, and their functionality is crucial for cell survival, especially under stress conditions (Cui et al, Fungal Genet Biol 103:42–54, 2017). In fungi, mitochondrial function can be assessed by monitoring the ability of yeast cells to grow on non-fermentable carbon sources (NFCs) such as glycerol or ethanol instead of carbon, which require functional mitochondria for energy production. This protocol describes the method for assessing mitochondrial health by growing yeast on agar plates with one of the NFCs instead of fermentable carbon. Cells with defective mitochondria show impaired growth on these NFCs, as they cannot efficiently generate energy through oxidative phosphorylation. In contrast, yeast with healthy mitochondria can grow well in the presence of these carbon sources. The protocol begins by growing yeast cells in a standard-rich medium to establish a starting population. The yeast cells are inoculated onto agar plates containing NFCs, such as glycerol, ethanol or lactate to assess their ability to utilize these substrates for growth (Smith et al, Endocr Rev 39:489–517, 2018; Zeng et al, Front Cell Infect Microbiol 13. https://doi.org/10.3389/fcimb.2023.1136698, 2023). The principle behind this assay is that NFCs require functioning mitochondrial respiration for energy production, as yeast cells can no longer rely on glycolysis alone. If the mitochondria are defective, the cells cannot efficiently metabolize these carbon sources and show poor or no growth on the plates. In contrast, wild-type yeasts with functioning mitochondria can grow vigorously. After inoculation, the plates are incubated at an optimal temperature (usually 30 °C) and the growth patterns are observed. The extent of growth is compared between different carbon sources and provides information about the health and functionality of the mitochondria. This method is widely used in mitochondrial research and genetic studies related to mitochondrial defects. This chapter provides a detailed protocol for testing mitochondrial function in yeast cells using

R. Pasrija, D. Kumari, *Protocols for Studying Pathogenic Fungi*,
https://doi.org/10.1007/978-981-96-7047-5_21

NFCs and provides a reliable, reproducible assay for mitochondrial health and viability.

Keywords

Fungi · Non-fermentable · Carbon sources · ATP · Mitochondrial · Fitness

Introduction This chapter describes protocols from media preparation to spot assays on plates with non-fermentable carbon sources (NFCs). Topics covered include:

21.1. Materials Required
21.2. Preparation of Media and Buffers
21.3. Protocol
21.4. Tips and Troubleshooting

21.1 Materials Required

21.1.1 Biological Materials

- **Fungal strains**
  - Pathogenic fungi such as *Candida* sp., *Saccharomyces* sp., *Cryptococcus* sp., *etc.*
  - Strains with mitochondrial defects or mutants in mitochondrial genes (optional) for comparative analysis.
- **Growth media**
  - **Minimal medium**: For the growth of fungi on non-fermentable carbon sources. This medium should not contain glucose but may contain the following: Yeast Nitrogen Base (YNB) or Synthetic Complete Medium (SC) without glucose and agar.
  - **NFCs:** Ethanol, glycerol or acetate (NFCs that fungi rely on for respiration, allowing the assessment of mitochondrial function).
  - **Fermentable carbon source (control):** Glucose (comparison of growth on fermentable versus NFCs).

21.1.2 Chemicals and Reagents

- **Buffered solution for growth:** Phosphate-buffered saline (PBS) (for washing fungal cells).
- **Agar (if plating):** Agar (1.5–2% w/v) (To solidify the medium when using plates).

- **Antibiotics (optional):** antimicrobial agents (*e.g.* penicillin, streptomycin) (To prevent bacterial contamination if required).

21.1.3 Laboratory Consumables

- **Petri dishes:** Sterile Petri dishes for plating fungal strains on agar media.
- **Centrifuge tubes:** Sterile tubes (15 or 50 mL) for the cultivation and harvesting of fungal cells.
- **Pipettes and tips** (Sterile pipettes and tips for liquid handling).
- **Microtiter plates (optional):** For high throughput analysis in liquid medium.

21.1.4 Equipment

- **Incubator:** For incubation of fungal cultures at 30 °C.
- **Centrifuge:** For pelleting fungal cells after growth, especially for measuring growth rates or for washing before inoculation.
- **Spectrophotometer (OD_{600}):** For measuring fungal growth using optical density (OD_{600}) to quantify cell proliferation.
- **Shaking incubator (when using liquid cultures):** To maintain fungal cultures in a liquid medium while shaking (200–250 rpm) for uniform growth.
- **Water bath (optional):** To control the temperature during the heating steps.

21.2 Preparation of Media and Buffers

21.2.1 Preparation of Yeast Extract Peptone Dextrose (YEPD) Media (See Chap. 2)

21.2.2 Preparation of PBS (See Chap. 3)

21.2.3 Preparation of 0.85% Saline (See Chap. 3)

21.2.4 Preparation of YNB with NFCs

- **Materials required:** YNB, agar, distilled water, glycerol (for a final concentration of 3%), ethanol (for a 2% final concentration), sodium acetate (for a 2% final concentration), glass beaker or flask, stirring bar, autoclave: for sterilization, sterile Petri dishes
- **Preparation protocol**
  - **Prepare the YNB solution**: Dissolve 6.7 g of YNB in approximately 900 mL of distilled water in a 1 L glass beaker or flask. Stir the solution thoroughly

until the YNB powder is completely dissolved. You may need to warm the solution slightly but do not boil it.

- **Add agar**: Add 20 g of agar to the YNB solution to solidify the medium. Mix thoroughly and make sure that the agar is completely dissolved. After mixing, adjust the final volume to 1 L with distilled water.
- **Add NFCs (anyone)**: Depending upon the choice of carbon, choose to add either 30 mL autoclaved glycerol (to achieve a final concentration of 3%), 20 mL ethanol (to reach a final concentration of 2%), or 20 g sodium acetate (to achieve a 2% final concentration) to the three different YNB dissolved earlier. Stir the ingredients gently and make sure that they dissolve completely. **Note:** After autoclaving, allow the medium to cool before adding 20 mL of ethanol.
- **Sterilize the medium**: Autoclave the prepared YNB medium at 121 °C for 15 min to sterilize it.
- **Pour agar plates**: Once the medium has cooled to the appropriate temperature (50–60 °C), pour the agar into sterile Petri dishes. Pour enough medium to cover the bottom of the Petri dish, but avoid overfilling. Allow the plates to solidify at room temperature. This usually takes 15–20 min.
- **Storage**: Once the YNB agar plates have solidified, they are ready for use. Store the plates at 4 °C in a sealed plastic bag or container to prevent dehydration and contamination. The plates can be stored for up to 2–3 weeks.

21.3 Protocol

- **Materials and reagents:** YNB agar plates supplemented with NFCs (*e.g.* glycerol, ethanol, sodium acetate) in place of carbon according to the protocol above, fungal culture, sterile PBS, saline, sterile pipette tips, sterile 1.5 mL microcentrifuge tubes or glass test tubes for serial dilution, spectrophotometer, an incubator set at 30 °C for fungal growth.
- **Protocol for spot assay on plates with NFCs**
  - **Preparation of the fungal suspension:** Grow the fungal culture in a liquid medium (*e.g.* YEPD medium) overnight at 30 °C in a rotary shaker (150 rpm). This enables exponential growth. Harvest the fungal cells by centrifuging the culture at 3000×*g* for 5 min. Discard the supernatant and wash the cell pellet twice with sterile PBS or distilled water. Resuspend the pellet in sterile PBS or water to obtain a cell density of approximately 10^6 cells/mL (OD_{600} ~ 0.1).
  - **Serial dilution:** Label five sterile microcentrifuge tubes as 10^{-1}, 10^{-2}, 10^{-3}, 10^{-4} and 10^{-5}. Add 900 μL of sterile saline to each tube. Mix 100 μL of the cell suspension with 900 μL of sterile saline in the first microcentrifuge tube (see Chap. 4). This results in a 1:10 dilution of the cells (approximately 10^6–10^7 cells/mL depending on the initial concentration). Take 100 μL of the cell suspension from the first tube (10^{-1}) and add it to the second tube (10^{-2}). Mix thoroughly. Repeat the procedure to prepare the subsequent dilutions. For example, take 100 μL from tube 10^{-2} and add it to the 10^{-3} tube, and so

on, until the 10^{-5} dilution is made. Finally, discard the 100 μL of cells from the fifth tube (see Chap. 4).

- **Spotting the fungal cells:** Using a sterile pipette tip, spot 5 μL of the fungal suspension onto the YNB agar plates containing the non-fermentable carbon sources. Typically, spot about 5–6 spots of serially diluted fungal suspension in a row per plate (Fig. 21.1), allowing enough space between the spots to form individual colonies. Alternatively, if testing different strains of fungi, we can spot each in different rows on the same plate. Like this, 3–4 strains can be tested in a single plate.
- **Incubation:** Incubate the plates at 30 °C for 2–3 days, depending on the type of fungus and its growth rate. The incubation time may vary (*candida/ Saccharomyces* sp.—48 h and *cryptococcus* sp.—72 h). Fast-growing strains may show visible growth within 24–48 days, while slower-growing strains may take 3 days to form colonies.
- **Observe and evaluate growth:** After the incubation period, observe the plates for the growth of fungal colonies at the site of the spot.
- **Record the results:**
  - **No growth**: indicates that the fungal strain is unable to utilize the NFCs, possibly due to mitochondrial defects or inability to metabolize the carbon source.
  - **Partial growth**: Indicates that the fungus can utilize the NFCs, but only to a limited extent, probably due to defects or slower metabolic activity.
  - **Complete/Full growth**: Indicates healthy mitochondrial function and the ability to utilize the NFCs.
- **Quantification (optional):** If necessary, quantify growth by counting the number of colonies in each spot. Compare growth between different spots to assess metabolic capacity with different carbon sources or fungal strains.

By following this procedure, spot assay can be used to assess the growth of fungi on NFCs, which gives insight into their metabolic preferences [1].

21.4 Tips and Troubleshooting

21.4.1 Tips

1. **Selecting the carbon source**
   - Select NFCs suitable for the fungal species to be analyzed. Common examples are:
     - Glycerol (used in mitochondrial respiratory pathways).
     - Ethanol (metabolized *via* the TCA cycle and oxidative phosphorylation).
     - Lactate or succinate (both are non-fermentable organic acids).
     - Acetate (can be metabolized via the citric acid cycle).
   - **Test multiple carbon sources:** For more comprehensive results, use different NFCs to assess mitochondrial fitness under different conditions.

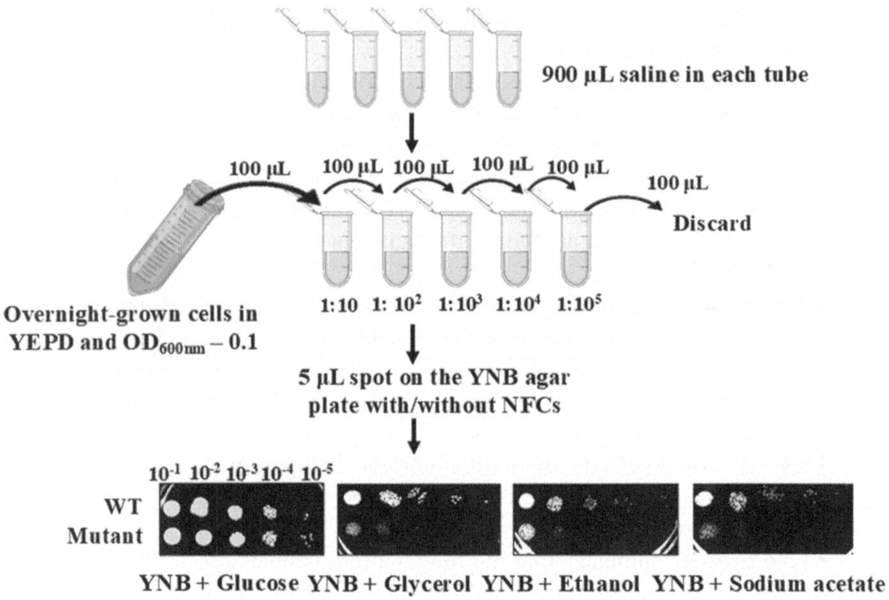

Fig. 21.1 Growth of WT and mitochondrial mutants of fungal cells on YNB + glucose (control) and plates with NFCs after incubation for 72 h at 30 °C

2. **Preparation of media**
   - **Minimal media with NFCs:**
     - Prepare a minimal growth medium (*e.g.* YNB to support fungal growth).
     - Supplement the medium with NFCs (*e.g.* 2% glycerol, 2% ethanol).
     - Yeast cells growing on glucose obtain their energy mainly through fermentation. So, the medium should not contain glucose or fermentable sugars, as these would allow growth by fermentation, bypassing mitochondrial respiration.
   - **Additional nutrients**
     - Ensure that the medium contains essential nutrients for mitochondrial function, such as ammonium salts (for nitrogen) and vitamins that support cellular metabolism.
3. **Conditions for fungal growth**
   - **Preparation of the culture**
     - First grow the fungi in a rich medium (*e.g.* YEPD) so that the cells reach the log phase.
     - Wash the fungal cells with sterile PBS or minimal medium before resuspending them in the experimental growth medium to remove residual nutrients that may interfere with the results.
   - **Preparation of the inoculum**

- Use an appropriate inoculum size (*e.g.* 10^6 cells/mL) to ensure clear growth differences, especially when comparing strains with different mitochondrial fitness.
- Ensure a uniform cell suspension in the growth medium to ensure a consistent inoculum density in all experiments.

4. **Incubation conditions**
   - **Temperature and time:**
     - Incubate the cultures at the optimal temperature for the fungal strain (usually 30 °C for most pathogenic fungi).
     - Monitor growth for an appropriate period (usually 72 h for *Cryptococcus* sp. and 48 h for *Candida*/*Saccharomyces* sp.), depending on the fungal species and growth rate.
   - **Shaking or static incubation**
     - For aerobic growth, it is usually recommended to incubate the cultures in a shaking incubator to ensure adequate oxygen supply.
     - Static incubation can be used to test anaerobic conditions (if relevant to the experiment).
5. **Monitoring and quantification of growth**
   - **CFU**
     - For more accurate quantification, grow serial dilutions on solid culture media and count CFUs to quantify viable cells (see Chap. 4).
   - **Microscopic observation**
     - Examine the morphology of the fungal cells under the microscope to look for signs of mitochondrial dysfunction, such as morphological changes in the fungal cells (*e.g.* reduced growth rate, abnormal shapes).

21.4.2 Troubleshooting

1. **Problem: Poor or no growth on non-fermentable carbon sources**
   - **Possible cause**: The fungal cells may not have sufficient mitochondrial function, contamination problems, or inappropriate experimental conditions.
   - **Solutions**:
     - **Check the concentration of the carbon source**: Ensure that the concentration of NFCs (*e.g.* glycerol or ethanol) is appropriate (usually 2–3% in the media). If the concentration is too low, the fungal cells may not be able to grow; if it is too high, they may be toxic.
     - **Check the media composition**: Check the media composition for all required components and ensure that the carbon source is properly dissolved and incorporated. Incorrect medium pH or missing components can inhibit growth.
     - **Inspect fungal viability**: Ensure the fungal strain is viable and not contaminated. If using a frozen strain, recultivate from a freshly revived strain

to rule out long-term storage issues that could affect mitochondrial function.
- **Use control strains**: Include wild-type or genetically characterized strains with known respiratory competence (*e.g.* able to grow on glycerol or ethanol) as positive controls. This ensures that the assay conditions are suitable for mitochondrial respiration testing.
- **Consider strain-specific growth conditions**: Different strains of fungi may have different sensitivities or growth requirements to NFCs. Refer to strain-specific protocols for any special growth requirements.

2. **Problem: Inconsistent spotting**
   - **Possible cause**: Uneven distribution of fungal cells during spotting or inconsistent inoculum density can cause uneven growth patterns.
   - **Solutions**:
     - **Standardize the cell density**: Standardize the fungal cell density in the inoculum using a spectrophotometer or a hemocytometer to measure cell concentration before spotting. Spot a fixed volume (*e.g.* 5–10 μL) for uniform inoculation.
     - **Use a sterile, consistent spotting technique**: Ensure that the spotting pipettes are sterile and that the fungal suspension is evenly applied to the plate. Use a sterile tip for each spot to avoid cross-contamination.
     - **Allow spots to dry**: Allow the spots to dry before incubation to avoid the spread of fungal cells, which can lead to uneven growth patterns or confluent growth.
3. **Problem: Fungal contamination**
   - **Possible cause**: Contamination from other fungal species or bacteria during handling or from the initial fungal culture.
   - **Solutions**:
     - **Use a sterile technique**: Always work in a sterile environment to avoid contamination. Sterilize tools (pipettes, tips) and media thoroughly before use.
     - **Check for contamination before spotting**: Check the purity of your fungal culture by examining it under a microscope for the presence of contaminating organisms (bacteria or other fungi).
     - **Use antibiotics if necessary**: If bacterial contamination is a problem, consider adding antibiotics (*e.g.* ampicillin or streptomycin) to the medium to reduce bacterial growth without affecting fungal growth.
4. **Problem: Incorrect temperature for incubation**
   - **Possible cause**: Incorrect incubation temperature can affect fungal growth ability, especially for strains that require specific temperature conditions for mitochondrial activity.
   - **Solutions**:
     - **Consider temperature sensitivity**: When working with respiratory-deficient mutants, ensure that growth conditions are not too harsh (*e.g.* avoid incubation at higher than optimal temperatures, which can exacerbate mitochondrial defects).

   - **Check the strain-specific temperature requirements**: In certain strains of fungi, mitochondrial respiration may be temperature sensitive, and growth on NFCs may be inhibited at higher temperatures.
5. **Problem: Mitochondrial defective strain not showing the expected phenotype**
   - **Possible cause**: The mutant may not have a complete mitochondrial defect or may have compensatory mechanisms. Alternatively, the mutations could be incomplete, resulting in a less pronounced defect.
   - **Solutions**:
     - **Confirm the mitochondrial mutation**: Make sure that the mutant strain has the expected mitochondrial defect (*e.g.* mutation of the respiratory chain or deletion of essential mitochondrial genes). Use PCR or sequencing to confirm the mutation.
     - **Test with alternative carbon sources**: Try additional NFCs (*e.g.*, ethanol, lactate) to ensure that mitochondria are defective and that the mutant cannot metabolize these substrates.
     - **Perform genetic or biochemical validation**: Confirm the mitochondrial defect by measuring mitochondrial functions, such as oxygen consumption (using a respirometer) or mitochondrial membrane potential in parallel experiments.
6. **Problem: Inconsistent results between plates**
   - **Possible cause**: Variations in media preparation, inoculum volume, or handling.
   - **Solutions**:
     - **Ensure consistency**: prepare all plates under the same conditions and use the same batch of media and fungal strains to minimize variability.
     - **Standardize the inoculum**: Ensure that the volume and concentration of the fungal inoculum are the same for all replicates to avoid discrepancies in growth patterns.
     - **Use image analysis**: When quantifying growth (*e.g.* colony diameter), use image analysis software to ensure consistent and reproducible data acquisition.

References

1. Coste A, Ramsdale M, Ischer F, Sanglard D (2008) Divergent functions of three *Candida albicans* zinc-cluster transcription factors (CTA4, ASG1 and CTF1) complementing pleiotropic drug resistance in *Saccharomyces cerevisiae*. Microbiol Read Engl 154:1491–1501. https://doi.org/10.1099/mic.0.2007/016063-0

In Vivo Assay Using *Caenorhabditis elegans* 22

Abstract

Caenorhabditis elegans, a small soil-dwelling nematode, has become one of the most valuable and widely used model organisms in biological and genomic research since its introduction into the laboratory over 50 years ago. Its appeal arises from its ease of maintenance, minimal nutritional requirements, rapid reproduction and long history of scientific study. *C. elegans* has the unique ability to self-fertilize and produce many offspring in a short period of time, with a life cycle of about 3 days at 25 °C. It was the first multicellular organism to have its genome fully sequenced, underpinning its importance in genomics. *C. elegans* has proven to be particularly useful for studies in genetics, cell biology, neurobiology and aging, thanks to its well-characterized genome and its transparent body, which allows non-invasive observation of internal structures. *C. elegans* has a simple but functional anatomy with a mouth, pharynx, intestine and gonads, with roughly 1000 somatic cells in the adult animal. Hermaphrodites and males differ in their sex chromosomes. Hermaphrodites have two X chromosomes, while males have only one X chromosome (XO). Their rapid growth, predictable developmental stages, and large offspring make them an ideal model for high-throughput research, including studies of aging, disease models, and drug testing (Stiernagle, WormBook. https://doi.org/10.1895/wormbook.1.101.1, 2006). This chapter provides a comprehensive guide to basic and essential protocols for handling *C. elegans*, making it a valuable resource for new and experienced researchers. Key techniques covered include media preparation, methods for synchronizing *C. elegans* cultures after embryo isolation, infection models, and microscopy techniques for observing nematode development. The chapter also provides guidelines for freezing and long-term preservation of strains. This chapter emphasizes the utility of *C. elegans* as a versatile and powerful organism in numerous areas of biological research. It provides a basic understanding of the use of this model in laboratory settings.

R. Pasrija, D. Kumari, *Protocols for Studying Pathogenic Fungi*,
https://doi.org/10.1007/978-981-96-7047-5_22

Keywords

C. elegans · Model organism · Protocols · Life cycle · Synchronization of cultures · Freezing

Introduction This chapter covers basic techniques for the maintenance of *C. elegans*. The following topics are covered as listed below:

22.1. Materials Required
22.2. Preparation of Media and Buffers for the Maintenance of *C. elegans*
22.3. Synchronization and Staging of the Cultures
22.4. Picking of *C. elegans*
22.5. Infection Model
22.6. Guidelines for Freezing Worms for Prolonged Storage
22.7. Tips and Troubleshooting

22.1 Material Required

22.1.1 Biological Materials

- **For C. elegans:**
  - ***C. elegans* strains**: Wild-type *C. elegans* (*e.g.* N2 Bristol strain) or transgenic strains (e.g., GFP-labelled or mutant strains) for easier monitoring.
  - **Nematode growth medium (NGM)**: Agar-based medium for cultivating *C. elegans*. This includes:
    - Agar, NaCl, peptone, cholesterol, and calcium chloride.
    - Additives (*e.g.,* ampicillin/streptomycin) to prevent bacterial contamination.
    - M9 buffer for washing and storing worms.
  - ***C. elegans* food source:** *Escherichia coli* OP50 (The standard food for *C. elegans* culture)
- **For the fungal culture**
  - **Fungal strains:** *Candida albicans/Cryptococcus neoformans* or any fungal strain.
  - **Fungal growth media:** Yeast extract peptone dextrose (YEPD) agar or sabouraud dextrose agar (SDA), brain-heart infusion (BHA) agar.
  - **Sterile PBS or phosphate-buffered saline (PBS):** For washing fungal cells before infection of *C. elegans*.

22.1.2 Chemicals and Reagents

- **Antibiotics (optional): Penicillin/streptomycin** (to prevent bacterial contamination during cultivation of the fungal strains)
- **Formaldehyde (optional for fixation):** For fixation of *C. elegans* after the assay for analysis under the microscope.

22.1.3 Laboratory Consumables

- **Petri dishes:** Sterile Petri dishes (90 mm) for cultivating *C. elegans* and *C. neoformans*.
- **Pipettes and sterile tips:** For accurately handling fungal suspensions, nematodes, and other solutions.
- **Sterile glass vials:** For storage of *C. elegans* or fungal cultures, if required.
- **Agar plates:** Culturing *C. elegans* (NGM agar plates) and *C. neoformans* (SDA or YEPD, BHI agar plates).

22.1.4 Equipment

- **Incubator:** For cultivating *C. elegans* (at 18–22 °C) and *C. neoformans* (at 25–30 °C).
- **Microscope:** A light or fluorescence microscope is used to observe the nematodes and *C. neoformans*.
- **Hemocytometer or spectrophotometer:** Used to count the concentration of *C. neoformans* cells before infection.
- **Stereomicroscope:** To observe the nematodes for motility and survival during the assay.
- **Timer:** To track the survival time of *C. elegans* after infection.

22.2 Preparation of Media and Buffers for the Maintenance of *C. elegans*

22.2.1 Preparation of NGM Plates

- **Materials required:** 1 M potassium phosphate buffer (KH_2PO_4) (pH 6.0), 5 mg/mL- cholesterol in ethanol, 1 M calcium chloride ($CaCl_2$), 1 M magnesium sulfate ($MgSO_4$), agar, peptone, sodium chloride (NaCl), 5 mg/mL 5-Fluoro-2′-deoxyuridine (FUdR) (optional, to prevent progeny), distilled water, *etc.*
- **Preparation steps:**
  - **Prepare the base medium**: Mix 1 L distilled water with 3 g NaCl, 2.5 g peptone and 17 g agar in a flask. Heat the mixture and bring to a boil to dissolve

all components. Autoclave the flask at 121 °C for 15 min to sterilize the medium.

- **Add buffers and salts**: After autoclaving, allow the medium to cool slightly (but not too cold so that the agar does not set). Once the temperature is around 60 °C, add the following sterile solutions in the order indicated: 25 mL of autoclaved 1 M KH_2PO_4 buffer (pH 6.0), 1 mL of filtered 1 M $MgSO_4$ and 1 mL of filtered 1 M $CaCl_2$. Mix the medium carefully to ensure an even distribution of the salts and buffer.
- **Add cholesterol and FUdR**: Add 1 mL of the filtered cholesterol solution (5 mg/mL in ethanol). Stir gently to ensure it is evenly distributed in the agar. If you want to prevent progeny production in your nematode culture, add 1 mL of FUdR solution (5 mg/mL) in this step. FUdR is commonly used to inhibit embryonic development, especially in experiments where you want to prevent population growth.
- **Pour plates**: Pour the prepared NGM mixture into sterile 90 mm petri dishes. Make sure that the depth of the medium does not exceed 6 mm (approximately 13 mL per plate) and allow the plates to cool and solidify at room temperature. Ensure that the plates do not dry out.
- **Storage**: Once the NGM has solidified, store the plates in a sterile environment (*e.g.* in a laminar flow hood or in a clean area) at 4 °C until use. The plates can be stored for several weeks under these conditions.

- **Notes:**
  - **Follow the sequence:** Follow the exact sequence of chemical additions as described in the protocol. This ensures proper dissolution and interaction of the ingredients. Adding certain salts and buffers in the wrong order can affect the pH of the medium or the solubility of other compounds. For example, adding the phosphate buffer before the salts ($MgSO_4$ and $CaCl_2$) will ensure proper buffering capacity and prevent precipitation. Always allow the medium to cool to around 60 °C before adding sensitive components such as cholesterol and FUdR.
  - **Sterility:** Ensure all chemicals added to the NGM are sterile, especially the cholesterol and FUdR solutions. If FUdR is added, work in a sterile environment to avoid contamination.
  - **Handling ethanol:** When adding cholesterol (dissolved in ethanol), ensure the ethanol has completely evaporated by keeping the solution at 60 °C or using a laminar flow hood. This will prevent ethanol residues from affecting the nematodes.
  - **Optional additives:** FUdR is not always required and is used to prevent progeny growth, e.g. when working with a specific nematode strain or in experiments where progeny would interfere with the results.
  - **Filter sterilized chemicals** (add after autoclaving):
    - **Cholesterol**: Cholesterol is heat-sensitive and should be filtered (typically 0.22 μm filter) and added after autoclaving when the agar mixture has cooled to about 60 °C.

- **$MgSO_4$ and $CaCl_2$:** These two are also heat-sensitive and should not be autoclaved.
- **FUdR**: FUdR is also heat-sensitive and should be filtered and added after autoclaving to avoid degradation during autoclaving.

22.2.2 Preparation of Bacterial Food Source

- **Materials required:** 300 mg KH_2PO_4, 600 mg disodium phosphate (Na_2HPO_4), 500 mg NaCl, 100 μL 1 M $MgSO_4$, 1 mL dextrose (1 g/5 mL of MQ), 200 μL uracil (2 mg/mL of MQ), 1 mL ammonium chloride (NH_4Cl) (1 M), *E. coli* OP50 culture, and water for 100 mL.
- **Preparation steps:**
  - **Prepare the base medium:** Dissolve 300 mg KH_2PO_4, 500 mg NaCl, and 600 mg Na_2HPO_4 in 100 mL distilled water in a clean container.
  - **Sterilize the medium:** Heat the mixture and boil to dissolve all components. Autoclave the flask at 121 °C for 15 min to sterilize the medium.
  - **Add remaining filtered sterilized compounds:** Add 100 μL of filtered 1 M MgSO4 to the solution and stir well. Add 1 mL of filtered dextrose, 200 μL of uracil and 1 mL of NH_4Cl. Mix well. Add 500 μL of the overnight grown culture of *E. coli* strain OP50 in LB media. Mix well and store overnight at 37 °C. Store the M9 buffer containing OP50 at 4 °C for later use.
- **Note:**
  - **Uracil exclusion**: Do not add uracil to the control flask as this prevents bacterial growth. This allows a comparison between the test flasks and the control by observing the turbidity. The flask with uracil should show bacterial growth, whereas the control does not, as no uracil was added.
  - **Storage:** The prepared bacterial food source can be stored in sterile containers at 4 °C and used for feedings of *C. elegans*, but it is recommended to use it within 1 week for optimal results.
  - **Sterilization:** Filter-sterilize sensitive chemicals like $MgSO_4$, dextrose, uracil, and NH_4Cl to ensure the bacterial growth medium is not contaminated.
  - **Turbidity measurement:** The turbidity of the bacterial culture is often used as an indicator of bacterial growth. If the control flask shows no turbidity while the experimental flask is turbid, this indicates proper exclusion of uracil in the control.

22.2.3 Preparation of M9 Buffer for *C. elegans*

- **Preparation steps:**
  - **Prepare the M9 buffer solution:** Add 3 g of KH_2PO_4, 6 g Na_2HPO_4 and 5 g NaCl to a clean, sterile container (*e.g.* a beaker or flask). Add 1 L of distilled water to the container and stir the mixture thoroughly to ensure complete dis-

solution of the salts. All components must be completely dissolved before proceeding to the next step.

- **Sterilize the solution:** After the salts are dissolved, sterilize the solution at 121 °C for 15 min. Autoclaving ensures that all microbial contaminants are removed from the buffer so that it is safe for use in experiments.
- **Add $MgSO_4$:** After autoclaving, allow the M9 buffer to cool to room temperature. Filter the 1 M $MgSO_4$ solution separately using a 0.22 μm filter to avoid contamination. Add 1 mL of filter-sterilized 1 M $MgSO_4$ solution to the cooled M9 buffer solution. Stir the solution gently to ensure the magnesium sulfate is well-mixed and evenly distributed in the buffer.
- **Storage of the M9 buffer:** The prepared M9 buffer can be stored at room temperature for short-term use or refrigerated at 4 °C for long-term storage. Ensure that the buffer is stored in a sterile container to avoid contamination. If kept refrigerated, it is best used within a few months.

22.2.4 Seeding NGM Plates with *E. coli* OP50

- **Ingredients:** NGM plates, *E. coli* OP50, sterile technical equipment (flame or ethanol, sterile pipettes, *etc.*).
- **Preparation steps:**
  - **Prepare the NGM plates and the *E. coli* OP50 culture:** Take the already prepared and sterilized NGM plates and make sure they are at room temperature. Prepare the *E. coli* OP50 culture in M9 buffer or LB medium. The culture should be grown fresh, usually overnight, to ensure the bacteria are in the exponential growth phase, which is critical for proper lawn formation.
  - **Pipetting the *E. coli* culture:** Using a sterile technique, pipette approximately 150–200 μL of the *E. coli* OP50 culture onto the surface of the NGM plates. This amount of culture is usually sufficient for a 90 mm plate. Spread the culture evenly over the surface of the NGM plate using a sterile glass spreader or a sterile pipette tip. To do this, carefully spread the liquid culture in a circular motion, ensuring it is evenly distributed over the surface without spreading too close to the edges.
  - **Dry the plates:** After spreading the bacterial culture, allow the plates to dry for a few minutes in a sterile environment, e.g. in a laminar flow hood, with the lids slightly open so that excess moisture can evaporate. This step is important to prevent the culture from becoming too wet, which could prevent the formation of a proper bacterial lawn.
  - **Incubation:** Once the plates are dry, place the colonized NGM plates in an incubator: For overnight incubation, incubate the plates at 22 °C to allow the bacterial lawn to grow slowly and evenly. For faster growth, incubate the plates at 37 °C for about 8 h. The plates should be kept at these temperatures until the bacterial lawn has formed, which takes 8–24 h depending on the incubation conditions.

- **Inspecting the bacterial lawn:** After the incubation time, check the plates to ensure that a continuous bacterial lawn has formed that is uniform and dense but does not spread too far toward the edges of the plate. Excess bacteria at the edges can cause the worms to crawl up the sides and possibly dry out or die.

- **Storage and use:** As soon as the bacterial lawn has formed, the plates can be used immediately to cultivate *C. elegans*. If the plates are not needed immediately, store them in a sterile, airtight container at 4 °C to prevent contamination and preserve the bacterial lawn. Seeded NGM plates can usually be stored at 4 °C for 2–3 weeks.
- **Notes:**
  - **Sterile technique:** Always use proper sterile technique throughout the process, especially when handling the *E. coli* OP50 culture and NGM plates, to avoid contamination.
  - **Avoid excessive spreading:** Make sure the bacterial lawn is confined to the center of the plate and does not spread to the edges. This will prevent the *C. elegans* from climbing up the sides of the plate, where conditions are less optimal.
  - **Incubation temperature:** Adapt the incubation conditions to your experimental requirements. While 22 °C is typical for keeping *C. elegans* in a laboratory, a higher temperature (37 °C) may promote faster bacterial growth.

22.2.5 Preparation of Luria-Bertani (LB) Broth

- **Materials required:** 25 g LB broth powder, 1 L distilled water, sterilized beakers or flasks, autoclave, *etc*.
- **Preparation steps:**
  - **Dissolve the LB broth**: Add 25 g of LB broth powder to 1 L of distilled water and stir until the powder is completely dissolved.
  - **Sterilize**: Autoclave the prepared LB broth at 121 °C for 15–20 min to sterilize it.
  - **Storage**: Allow the medium to cool before use and store the sterilized LB broth at room temperature.

22.2.6 Preparation of BHI Agar Plates

- **Ingredients:** BHI agar powder, ampicillin (100 mg/mL stock solution), distilled water, sterilized flask, autoclave, *etc*.
- **Preparation steps:**
  - **Prepare BHI agar**: Weigh out 37 g of BHI agar powder. Add the 37 g of BHI agar to 1 L of distilled water. Stir the mixture well so that the agar powder dissolves completely.

- **Sterilization**: Autoclave the BHI agar mixture at 121 °C for 15–20 min to sterilize it. Once autoclaving is complete, allow the medium to cool to approximately 50–60 °C in a sterile environment.
- **Add ampicillin**: Prepare an ampicillin stock (usually 100 mg/mL in water). Add 100 μL of 100 mg/mL ampicillin to the cooled BHI agar (100 mL) (final concentration 100 μg/mL). Mix gently to avoid bubbles and to ensure an even distribution of ampicillin throughout the agar.
- **Pour the plates**: Pour the cooled, ampicillin-supplemented BHI agar of 4–6 mm deep into sterile 90 mm Petri dishes (approximately 12–15 mL per plate) and allow the plates to set at room temperature, making sure that they are flat.
- **Storage**: Store the prepared plates at 4 °C to maintain sterility. Plates can be used within 1–2 weeks for fungal infection assays with *C. elegans*.

• **Notes:**
  - Ampicillin prevents bacterial contamination, particularly if bacterial growth could interfere with fungal infection or *C. elegans* assays.
  - Ensure the BHI agar is freshly prepared or properly stored to maintain its nutritional quality. When working with ampicillin, always handle the antibiotic with care, as it may degrade with prolonged exposure to light or heat.

22.2.7 Preparation of YEPD Media (See Chap. 2)

22.3 Synchronization and Staging of the Cultures

22.3.1 Transferring Worms Grown on NGM Plates

Various methods exist for transferring *C. elegans* from one Petri dish to another. A quick and convenient method is "chunking", where a sterilized scalpel or spatula transfers a piece of agar from an old plate to a new plate.

• **Materials required**
• NGM plates, *E. coli* OP50, sterile technique equipment (flame or ethanol, sterile pipettes, coverslips), *etc.*
• **Protocol**
  - Ensure the initial plates contain healthy worms and a lawn of *E. coli* OP50.
  - Using a sterile technique, flame, or ethanol, sterilize a forceps and coverslip. Carefully cut off a piece of agar containing worms along with the OP50 bacterial lawn from the source NGM plate. The piece should contain a considerable number of worms.
  - Immediately transfer the piece of worms to the surface of the target NGM plate containing the OP50 bacteria.
  - Allow the worms to acclimatize to the new plate for a few minutes.

- Once the transfer is complete, discard the source NGM plate or store it if desired. Keep the plates at 22 °C for 2–3 days to obtain adult worms.

22.3.2 Embryo Isolation

- **Materials required**
- Sodium hypochlorite solution (household bleach diluted with water to a final concentration of 4%), 5 mL of 1 M sodium hydroxide (NaOH) solution, M9 buffer for washing the worms and resuspending the embryos, worm pin or platinum wire for transferring the worms, for sterile equipment, flame or ethanol, sterile microcentrifuge tubes, sterile pipette tips and pipettes, NGM seeded with *E. coli* OP50 seeded NGM plates for plating the embryos, centrifuge with 2000–2500 rpm.
- **Protocol**
  - **Prepare the bleach solution:** Dilute household bleach (sodium hypochlorite) with water to make a 4% bleach solution. This is used to dissolve the cuticle of the adult worms and release the embryos.
  - **Labelling:** Label a sterile 15 mL tube for each sample to be processed. This contains the isolated embryos.
  - **Collect the worms:** Take 3–4 source plates with adult *C. elegans* nematodes. Wash the worms off the plates by adding M9 buffer (1–2 mL) to the plate and gently shaking the plate to dislodge the worms. Transfer the worm-buffer solution to the next plate to collect more worms and continue this process for all plates. Combine the collected worms and M9 buffer into a 15 mL conical tube.
  - **Centrifuge and wash the worms:** Centrifuge the 15 mL tube at 2000 rpm for 1.5 min to pellet the worms and carefully discard the supernatant without damaging the pellet. This will ensure that the worms are free of bacteria and contaminants.
  - **Prepare the bleach solution:** After washing, add 5 mL of 1 M NaOH and 2 mL of 4% bleach solution to the tube containing the worm pellet and mix gently by inverting the tube a few times. Remove the cap from the tube to allow the bleach solution to act effectively on the worms.
  - **Incubate to release the embryos:** Allow the solution to incubate for 2–3 min at room temperature. 80–90% of the adult worms will dissolve during this time and release the embryos. The worms will appear as tiny dots under the microscope (Fig. 22.1).
  - **Centrifuge to pellet the embryos:** Centrifuge the tube after 2–3 min at 2500 rpm for 1 min to pellet the embryos. Discard the supernatant and wash the pellet 2–3 more times with M9 buffer to remove all residual bleach and NaOH.
  - **Resuspend the embryos:** Leave a few mL of M9 buffer in the tube to resuspend the embryos. This amount can be adjusted depending on the final volume needed for plating.

Fig. 22.1 Embryo isolation

  - **Plate the embryos:** Transfer 200 μL of the embryo suspension to a freshly seeded NGM plate with *E. coli* OP50. Spread the embryos evenly on the plate using a sterile pipette tip or worm pick. Incubate the plates at 22 °C for 48 h.
  - **Check for the L4 stage:** After 48 h of incubation, check the plates for the appearance of the L4 stage larvae. This confirms the successful isolation and development of the embryos.
- **Notes:**
  - The bleach solution should be used with caution. Ensure adequate ventilation as the vapours from the bleach can be harmful.
  - When centrifuging, ensure the embryos are not disturbed or excessively pelleted.
  - To ensure optimal isolation of the embryos, NaOH and bleach treatment should be given carefully in a time-bound manner. Excessive exposure can damage the embryos or impair their viability.
  - The washing steps with the M9 buffer will help to remove bacteria and other contaminants from the worm population, leaving you with clean embryos for further experimentation.
  - When spreading the embryos, do not allow the plate to dry out completely as this may hinder larval development. Only allow the plate to partially dry (semi-dry) before further incubation. This ensures optimal conditions for embryo development.
- **Identification of the L4 stage:** L4 larvae can be identified by their characteristic shape and features under the microscope:
  - **Tail:** L4 larvae have a distinct, pointed tail and a more developed gonad than earlier stages.
  - **Cuticle:** The cuticle is more robust, with visible segmentation and a clearly defined L4 pattern.
  - **Locomotion:** The L4 larvae are generally more active than earlier larvae and show a typical sinusoidal movement.
  - **Appearance of the vulva:** The vulva of the L4 stage appears as a distinct semicircle on the ventral side of the larva. This is a distinguishing feature, as it is more visible in the L4 stage than in earlier stages.

22.3.3 Life Cycle from Egg to Egg-Bearing Adult

Embryogenesis of *C. elegans* takes about 16 h at 20 °C. After fertilization, an impermeable egg membrane forms, which enables independent embryonic development. The embryos remain in the hermaphrodite until about the 24-cell stage before they are laid. After hatching, the hermaphrodite embryo has 558 cell nuclei and develops into the L1 larval stage. The larvae go through four stages (L1–L4), with L1 lasting about 12 h and the subsequent stages lasting about 8–9 h each. Each stage ends with lethargus, a period of inactivity to renew the cuticle. Adult hermaphrodites start reproducing about 18 h after the L4 molt and produce offspring for 2–3 days until the self-produced sperm are exhausted. After reaching the L2 stage, the organism may enter a dormant state known as dauer larva if environmental conditions are unfavorable for further growth. In this dauer stage, feeding ceases indefinitely and locomotion is considerably restricted. Exiting the dauer stage occurs immediately as soon as favorable conditions prevail. The organism resumes feeding within an hour of gaining access to food. Feeding begins within 2–3 h, followed by molting into the L4 stage about 10 h later. Further offspring is possible when the hermaphrodite mates. After reproduction, hermaphrodites live for several weeks until senescence [5]. The complete life cycle of the WT Bristol N2 worm is illustrated in Fig. 22.2.

22.4 Picking out *C. elegans*

- **Materials required:** Source NGM plate with *C. elegans*, target NGM plate, sterile worm pick or platinum wire, sterile technical equipment (flame or ethanol, dissecting microscope, *etc.*)
- **Protocol**
  - Prepare both the source and target NGM plate in advance. Ensure that the target plate is labeled accordingly.
  - Sterilize the worm pick or platinum wire by holding it in a flame or dipping it in ethanol and then flaming it. Allow it to cool briefly to avoid damaging the worms.
  - Using a dissecting microscope, locate the area of the source NGM plate containing the worms you wish to transfer.
  - Insert the sterilized worm stick or platinum wire into the bacterial lawn near the worms. Avoid ingesting large amounts of bacteria.
  - Gently scrape the surface of the NGM plate with the worm pick or platinum wire to detach the worms from the plate (Fig. 22.3). Be careful not to damage the worms or disturb the bacterial lawn too much.
  - Transfer the worms from the source plate to the target plate by gently holding the tip of the worming needle or platinum wire to the surface of the target plate where you want to deposit the worms (shown in Fig. 22.3).
  - Gently release the worms onto the target plate by tapping the worm pick or platinum wire against the agar surface. Ensure that the worms are evenly distributed on the plate if necessary.

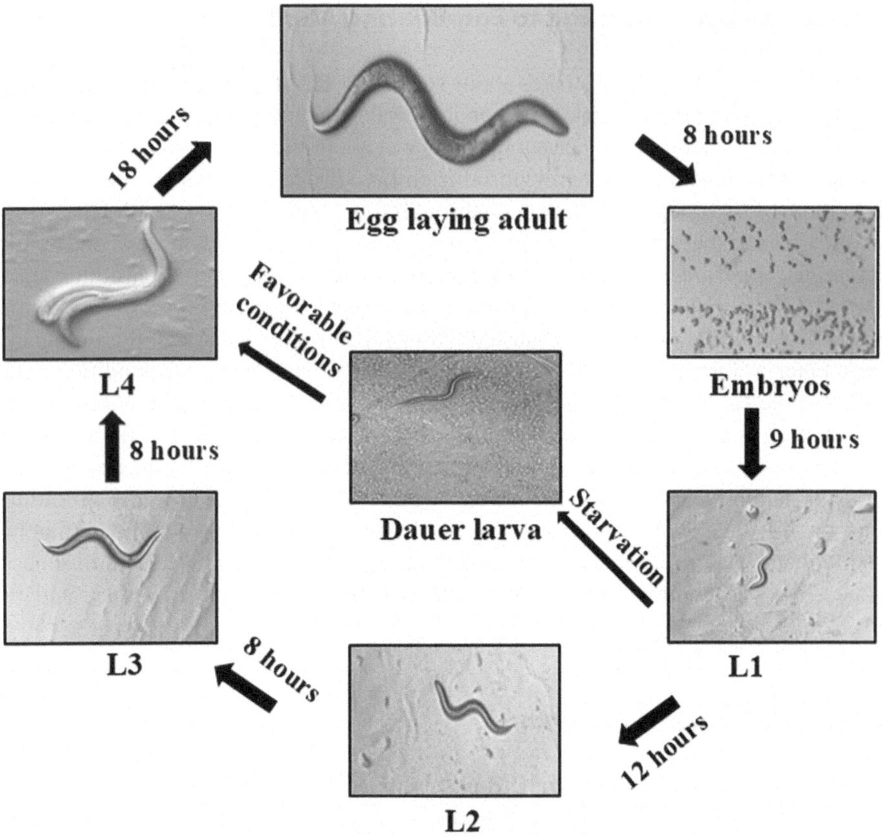

Fig. 22.2 Life cycle of *C. elegans*. (The photos were taken on Petri dishes under 10× magnification)

- Repeat steps 4–7 to transfer all the worms from the source plate to the target plate.
- Once you have transferred all the worms, dispose of the source plate or save it for later use if desired.
- Allow the worms to acclimate to the new plate for a few minutes before continuing the experiments.

22.5 Infection Model [2]

- **Materials required:** *C. elegans* nematodes in the L4 stage, NGM-OP50 plates with *E. coli* OP50 as a food source for the worms, BHI agar plates: for fungal growth, fetal bovine serum (FBS) to induce hyphal formation in *Candida* on BHI

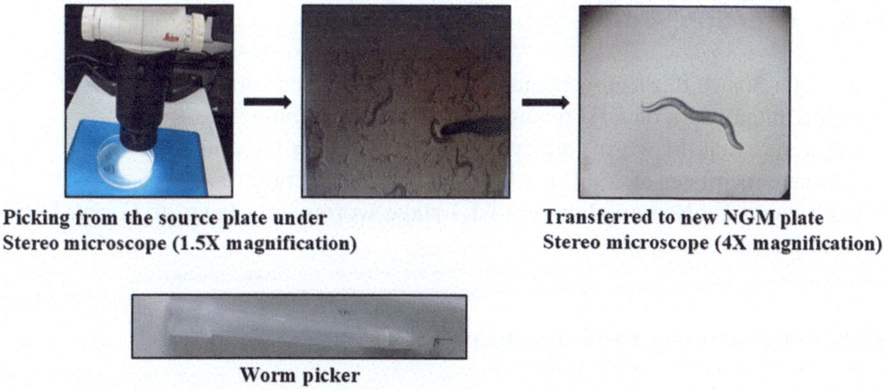

Fig. 22.3 Picking of worms to new NGM plate

plates, *Candida* sp. or *Cryptococcus* sp. fungal pathogen for the infection model, M9 buffer for washing the worms and preparing fungal suspensions, platinum wire for transferring the worms and touch test, light microscope to study the interactions between worms and fungi, sodium azide (NaN_3) to immobilize the worms during photography, ampicillin to inhibit bacterial contamination in the fungal culture and an incubator set at 25 °C for worm and fungal growth [1], GraphPad Prism for statistical analysis of survival data, *etc*.

22.5.1 Preparation of Fungal Culture

- Collect *C. albicans* cells from YEPD plates (or use other fungi such as *Cryptococcus* sp. if testing with them).
- Wash the yeast cells twice with M9 buffer by resuspending them in 1 mL of M9, then centrifuge at 1000 rpm for 1 min and discard the supernatant.
- Adjust the cell concentration to 5×10^7 cells/mL in M9 buffer.
- Spread 100 μL of the prepared yeast suspension (5×10^7 cells/mL) on BHI agar plates.
- Add ampicillin (100 μg/mL) to the plates to inhibit bacterial growth.
- Allow the plates to grow overnight at 25 °C so the yeast cells can form a lawn.
- **Observe hyphal formation:** To induce hyphal formation in *Candida* cells, add FBS to the BHI plates usually at a concentration of 10%. The hyphal formation is critical for mimicking the pathogenic form of *Candida* in *C. elegans* infection models.

22.5.2 Infection Setup

- Transfer 30–40 *C. elegans* nematodes at the L4 stage from an NGM-OP50 plate to the fungal lawn on BHI plates using a sterile platinum wire.
- Make sure that the worms are evenly distributed over the fungal lawn.
- Incubate the plates at 25 °C to allow the infection to progress.
- As a control, place another set of L4 stage worms on a lawn of *E. coli* OP50 (uninfected worms) [3].

22.5.3 Monitoring the Infection

- Count the number of live and dead worms every 24 h.
- Dead worms do not respond to contact with a platinum wire pick (a gentle nudge should cause the worm to move).
- Transfer worms to fresh plates every 24 h to prevent the accumulation of offspring, as this could falsify the results.
- **Control group:** Worms kept on NGM-OP50 plates (without fungal infection) should be used as controls to compare the survival rate with infected worms.

22.5.4 Imaging the Worms

- For imaging, immobilize the worms by placing them on a 2% agarose pad (as shown in Fig. 22.4).
- Add 5 μL of 30 mM NaN_3 in M9 buffer to the worms to induce paralysis.
- Use a light microscope to photograph the immobilized worms and take pictures of the fungal hyphae/fungal load interacting with the worms

22.5.5 To Identify Gastrointestinal Abnormalities

Healthy worms grown on NGM-OP50 plates have a well-defined, straight intestinal tract and a functional, regular pharynx (Fig. 22.4, right side (40×)). However, fungal-infected worms may show significant changes, such as intestinal damage or pharyngeal swelling, which can indicate fungal virulence.

- **Abnormalities in the pharynx**: Infected worms may show altered pharyngeal structure due to fungal infection, such as swelling, deformation or lack of normal pumping activity. The pharynx can be examined under a 10× to 100× microscope to observe changes in form or function.
- **Abnormalities of the intestine**: Infected worms may have distorted or damaged intestines, which may be swollen and filled with fungal hyphae (Fig. 22.4, left side (10×)) or show signs of obstruction. The intestine can be easily visualized

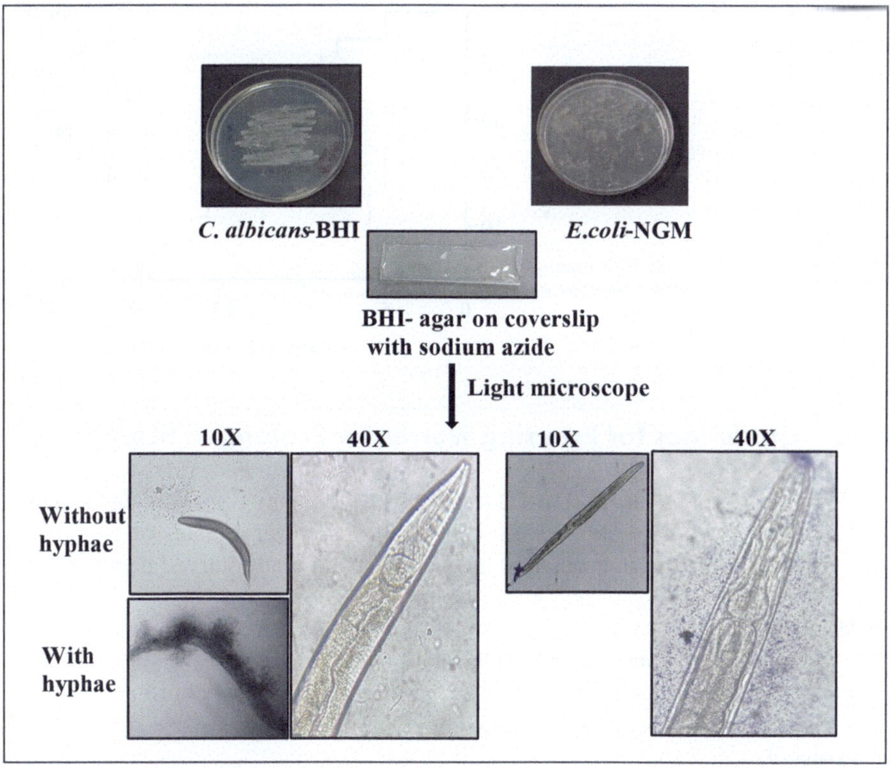

Fig. 22.4 Bristol N2 worms were fed either *C. albicans* (under hyphae-inducing conditions or not) or *E. coli* OP50 (control). After infection, the *C. elegans* worms were collected. The image shows hyphae emerging from the worms (10× magnification), and a distorted pharynx and intestine in worms fed with *Candida* cells (40× magnification), compared to a healthy worm from the control group (right)

under the same magnification range (10× to 100×) by looking for irregularities in the intestinal lumen or structure.

22.5.6 Data Analysis

- Plot the survival curves of the infected and control worms as shown in Fig. 22.5.
- The LT50 (lethal time in which 50% of the worms die) can be calculated using the Kaplan–Meier method.
- Statistical analysis should be performed using the log-rank or Wilcoxon tests to determine significant differences between the experimental and control groups. p values <0.05 are considered statistically significant [4].

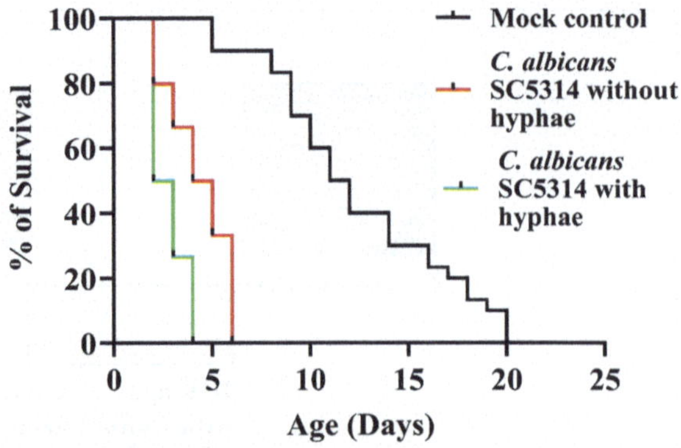

Fig. 22.5 Kaplan-Meier curve of percentage survival of worms fed *E. coli* OP50 (control) and *C. albicans* SC5314 strains under hyphae-inducing and non-hyphae-inducing conditions

22.6 Guidelines for Freezing Worms for Prolonged Storage

Freshly starved young larvae (L1–L2 stage) survive freezing best. Well-fed animals, adults, and eggs do not survive well. It is best to use several plates of worms that have just exhausted the *E. coli* OP50 lawn and contain many starved dauer larvae.

- **Materials**
  - Source NGM plate with 13–15 days old *C. elegans*
  - Cryogenic tubes, 0.08 M trehalose, 0.5 M DMSO, and M9 buffer
  - Sterile technical equipment (flame or ethanol, *etc.*)
- **Protocol**
- Prepare a freezing solution with trehalose and DMSO. To prepare a 50 mL freezing solution, add 1.51 g trehalose and 1.78 mL DMSO and mix in 50 mL M9 buffer.
- Collect *C. elegans* worms at the dauer stage (after 13–15 days of starvation in NGM-OP50 plate) with M9 buffer (mix worms from 3 plates to a 15 mL Falcon tube).
- Centrifuge at 2000 rpm for 1.5 min and remove excess buffer. Repeat this procedure 3 times.
- Add 1 mL of freezing solution to the tube and centrifuge at 2000 rpm for 1.5 min and remove excess solution
- Add 3 mL freezing solution and keep at RT for 45 min.
- Transfer the solution to 3 cryotubes (1 mL in each tube), store the tubes in a cardboard box and store at −80 °C for 6 months.

22.7 Tips and Troubleshooting

1. **Problem: Low infection rate**
   - **Possible potential issue**: If there is a low infection rate (*i.e.*, only a few worms die from the fungal infection), this could be due to poor fungal growth or insufficient inoculum.

- **Solution**: Ensure that the fungal lawn is fresh and thick. The yeast suspension should be well-distributed on the BHI plate. If using *C.* albicans, ensure that FBS is included to promote hyphal growth.

2. **Problem: fungal growth is insufficient and contamination**
   - **Possible potential issue**: If the fungal growth is insufficient or the yeast is not forming the lawn, there could be problems with the viability of the fungi or the culture conditions.
   - **Solution**: Check the fungal culture for contamination. Ensure the BHI plates are enriched with ampicillin to inhibit bacterial growth and provide the necessary nutrients to the fungus. Provide the right temperature and time for the fungus to grow.
3. **Problem: Thick fungal growth lawn**
   - **Potential issue**: If the worms are inside the fungal lawn and you cannot touch them with the platinum wire to check their viability, it can be difficult to judge whether they are dead or alive.
   - **Solution**: Apply light pressure to the fungal lawn near the worm under the microscope. This may cause the worm to move out of the lawn, making it easier to check whether it is responding to touch. If the worm does not react after 4–5 touches times, one can safely assume that it is dead.
4. **Problem: Formation of offspring**
   - **Potential problem**: The formation of offspring can make it difficult to accurately count dead worms and analyze survival.
   - **Solution**: To prevent the formation of offspring, transfer the worms to fresh BHI plates every 24 h. This ensures that no offspring accumulate. This makes it easier to count the live and dead worms and avoid disturbance from young larvae.
5. **Problem: Counting the dead worms**
   - **Potential problem:** Difficulty counting dead worms if they do not respond to touch or movement.
   - **Solution**: If a worm does not respond to a gentle touch with a platinum wire 4–5 times, consider it dead. Also, transfer worms to fresh plates regularly to avoid confusion with offspring.
6. **Problem: Imaging the fungal load**
   - **Potential problem**: The fungal load may not be sufficiently visible in the worms if imaging is performed too early.
   - **Solution**: To obtain clear images of the fungal burden in the worms, wait 2–3 days before imaging. By this time, the fungal infection should be sufficiently established in the worms and hyphal growth inside the worms should be visible when viewed under the microscope.
7. **Problem: Low worm movement or poor worm health**
   - **Potential problem**: If the worms have little movement or look unhealthy, they may not be viable or suffer from infection stress
   - **Solution**: Ensure the worms have reached the L4 stage and are healthy before starting the experiment. Consider using a new batch of worms if signs of health problems are observed at the start of the experiment.

8. **Problem: Difficulty identifying fungal hyphae**
   - **Potential problem**: If fungal hyphae are not visible on microscopic images, the fungal cells (*Candida*) may not have fully transitioned to the hyphal form.
   - **Solution**: Ensure that FBS is included in the BHI plates for optimal hyphal induction. Additionally, check the growth conditions and consider adding a small amount of glucose to the BHI plates if fungal growth appears inhibited.

References

1. Desalermos A, Tan X, Rajamuthiah R, Arvanitis M, Wang Y, Li D, Kourkoumpetis TK, Fuchs BB, Mylonakis E (2015) A multi-host approach for the systematic analysis of virulence factors in *Cryptococcus neoformans*. J Infect Dis 211(2):298–305. https://doi.org/10.1093/infdis/jiu441
2. Mylonakis E, Ausubel FM, Perfect JR, Heitman J, Calderwood SB (2002) Killing of *Caenorhabditis elegans* by *Cryptococcus neoformans* as a model of yeast pathogenesis. Proc Natl Acad Sci 99(24):15675–15680. https://doi.org/10.1073/pnas.232568599
3. Mylonakis E, Idnurm A, Moreno R, El Khoury J, Rottman JB, Ausubel FM, Heitman J, Calderwood SB (2004) *Cryptococcus neoformans* Kin1 protein kinase homologue, identified through a *Caenorhabditis elegans* screen, promotes virulence in mammals. Mol Microbiol 54(2):407–419. https://doi.org/10.1111/j.1365-2958.2004.04310.x
4. Park H-EH, Jung Y, Lee S-JV (2017) Survival assays using *Caenorhabditis elegans*. Mol Cells 40(2):90–99. https://doi.org/10.14348/molcells.2017.0017
5. Carranza-García E, Navarro RE (2020) Insights into the hypometabolic stage caused by prolonged starvation in L4-adult Caenorhabditis elegans hermaphrodites. Front. Cell Dev. Biol. 8. https://doi.org/10.3389/fcell.2020.00124